D0200811

HOW
MEN
AGE

HOW MEN AGE

What EVOLUTION *Reveals about* MALE HEALTH *and* MORTALITY

RICHARD G. BRIBIESCAS

Princeton University Press
Princeton and Oxford

Published by Princeton University Press,
41 William Street, Princeton, New Jersey 08540
In the United Kingdom: Princeton University Press,
6 Oxford Street, Woodstock, Oxfordshire OX20 1TR
press.princeton.edu

Jacket images: left half courtesy of Getty Images; right half courtesy of
Thinkstock

Library of Congress Cataloging-in-Publication Data

Names: Bribiescas, Richard G., author.
Title: How men age : what evolution reveals about male health and mortality / Richard G. Bribiescas.
Description: Princeton, New Jersey : Princeton University Press, [2016] | Includes bibliographical references and index.
Identifiers: LCCN 2016014947 | ISBN 9780691160634 (hardback : alk. paper)
Subjects: | MESH: Men's Health | Aging | Men—psychology | Biological Evolution | Mortality
Classification: LCC RA777.8 | NLM WA 306 | DDC 613/.04234—dc23 LC record available at https://lccn.loc.gov/2016014947

British Library Cataloging-in-Publication Data is available

This book has been composed in Adobe Garamond Pro and
Trade Gothic LT Std

Printed on acid-free paper. ∞

Printed in the United States of America

10 9 8 7 6 5 4 3 2 1

To my wife, Audrey.
I am fortunate and blessed to grow old with you.

CONTENTS

ACKNOWLEDGMENTS

This project is a follow-up to my first book, published in 2006, *Men: Evolutionary and Life History*. Upon completion of that work, I felt each chapter could easily be developed into a book of its own. Among the topics that I wished to tackle in greater depth, aging was at the top of the list. The growing amount of gray in my beard and hair likely played a role in stoking my interest. Friends and colleagues encouraged me to pursue this project from its first inception. I would like to thank my editor at Princeton University Press, Alison Kalett. From our first conversation over coffee at the Human Biology Association meetings a few years ago to our discussions in my office at Yale in 2013, she has been a staunch advocate and supporter of this work. Her guidance and encouragement have been indispensable.

Among those who provided deserved and constructive criticism as well as support are Brenda Bradley, Mark Eggerman, Eduardo Fernandez-Duque, Andrew Hill, Marcia Inhorn, Catherine Panter-Brick, Alison Richard, Eric Sargis, Claudia Valeggia, and Brian Wood. Ph.D. students, past and present, have been constant sources of inspiration and energy. Among those are Dorsa Amir, Stephanie Anestis, Erin Burke, Kate Clancy, Jessica Minor, Michael Muehlenbein, and Angélica Torres. Other Ph.D. students and scholars who have made contributions through their dedication and research in the Yale Reproductive Ecology Laboratory include Gary Aronsen, Kendall Arslanian, Melanie Beuerlein, Sofia Carrera, Stephen Chester, Jessamy Doman, Amara Frumkin, Angie Jaimez, Max Lambert, Kristen McLean, Leon Noel, Aalyia Saddrudin, Amelia Sancilio, and Kyle Wiley.

Many have kept me from sailing into intellectually choppy waters or have been at the ready to scrape me off the rocky shore when I've crashed and floundered; these include Peter Ellison, Irven DeVore, Stephen Stearns, Grazyna Jasienska, Michal Jasienski, Anna Ziomkiewicz-

Wichary, Szymone Wichary, Akiko Uchida, Meredith Reiches, Herman Pontzer, Steve McGarvey, James Holland Jones, Magdalena Hurtado, and Peter Gray. Indeed, Jasienska has been instrumental in sharpening my thinking on trade-offs between reproduction and aging as well as being a tremendous friend. Shripad "Tulja" Tuljapurkar and his coauthors left an indelible impression on my thinking with their seminal paper in 2007 titled "Why Men Matter." Peter Ellison has been a steadfast touchstone for advice and intellectual guidance. Kim Hill introduced me to research among the Ache of Paraguay, and for those experiences I will always be thankful. Akiko was a most gracious host for my wife, Audrey, and me during our visit to Japan to study male aging.

I have been very fortunate to be welcomed into the Shuar Life History Project, making many new friends and learning much along the way. They include Larry Sugiyama, Josh Snodgrass, and Felicia Madimenos. Student researchers have been equally awesome and welcoming, including Tara Capon, Alese Colehour, Theresa Gildner, Chris Harrington, Melissa Leibert, Heather Shattuck-Heidorn, Julia Ridgeway-Díaz, and Sam Urlacher. In Ecuador, there are many to be thanked, including Oswaldo Mankash, Luzmilla, Estella, and Rosalinda Charo Jempekat, Don Guimo, Bertha Fernandez, Judith and Rosa at the Posada del Maple, Marcia Salinas, and Dr. Jaime Guevara-Aguirre. My former chair's assistant, Ann Minton, firewalled my time so I could squirrel away a few minutes here and there to think and write. My current assistant, Tracy Edwards, has taken up this torch and has been terrific in carving out time in my schedule to write. I am also grateful for the constructive comments of two anonymous reviewers.

Those not yet mentioned but who certainly merit a tip of the hat are Jim Saiers, Lynn Saiers, Marty Wallace, Janice Murphy-Wallace, Christopher Wallace, Gabriel Olszewski, Roger Ngim, and the entire administrative staff in Yale Anthropology as well as my friends and colleagues in the Yale Office of the Provost. Thanks for adding laughter, warmth, and friendship to my own journey of aging.

To my parents and family, especially my sister Loli, I am always grateful for your unflinching enthusiasm and for keeping me grounded in reality. Finally and foremost, I thank Audrey, my wife of over twenty-five years. Thank you for guiding and tolerating me through my own journey of aging with grace and humor. I hope that I have become a bit wiser, kinder, and easier to live with along the way.

Hamden, Connecticut
Sucúa, Ecuador
December 2015

HOW MEN AGE

CHAPTER 1

A GRAY EVOLUTIONARY LENS

Old age ain't no place for sissies.

—Bette Davis

I vaguely remember my father having black hair. Being the youngest of four and born to older parents, I recall my most vivid memories to be of him having silver hair and mustache. He was sort of a mix between Clark Gable and Cesar Romero. Lucky Mom. However, as he got older, it wasn't difficult to see changes in the way he looked, spoke, moved, and acted. Except for mild diabetes and minor back issues resulting from a tumble off a ladder, he sported relatively good health for most of his life. By the time I started attending UCLA as an undergraduate, I noticed that some of his tastes and habits had begun to change. Instead of having a beer or two with my uncles, he opted for coffee. He passed on the late-night movie, went to bed a bit earlier, and napped more often. Although he was a modest smoker in his younger years, I don't remember him picking up a cigarette after the age of forty or so. In his later years, he spent much of his time working with my older brother in their workshop. As he occupied himself with one craft project or another, I saw that he was starting to hunch slightly while he worked, as if he was in a slow but constant struggle against gravity. In his sixties, he was much thinner compared to his earlier cinematic appearances in family home movies. I suspect much of his weight control was the result of my mother's close attention to his diet because of his diabetes. However, I am certain he was simply declining as a result of aging.

Charles Darwin and my father both lived to the age of seventy-three. Both married, fathered children, had their share of health issues, and were outlived by their wives. While my father did not write any books, venture to the Galápagos, or have the fame of Darwin, both ultimately succumbed to the effects of aging and died of heart failure. Though my father's passing was sudden, his death was not shocking. He was a great husband, father, and grandfather. As he aged, his life changed and he adjusted with dignity and a wry sense of humor that was often capped by a rolling of the eyes and a smile. Though he was not a biologist and not very familiar with physiology, hormones, or the biological bits and bobs of aging, I am certain that during his later years he understood that he was not the man he had been when he was twenty. Seventy-three trips around the sun is an accomplishment for most men around the world. As with Darwin, Pop lived a long, rich life and was loved by many. By any measure, evolutionary or otherwise, my father was a success.

Thousands of miles away in South America, other men in the forests of eastern Paraguay were living parallel lives. Cuategi is a bit younger than my father but not by much. As a man from the Ache people, who until the 1960s and 1970s were full-time nomadic foragers with very limited contact with the outside developed world, he spent most of his time searching for food, keeping his family safe, and coping with the daily microdramas that are commonplace in all small communities. He made and stoked fires from damp wood, drank wild yerba mate from the *Ilex paraguariensis* bush (now widely available in the organic food section of your local supermarket), fashioned archery bows from palm wood using a modified land snail shell as a hand planer, and devoted many days to the pursuit of capuchin monkeys (*Cebus capucinus*), white-lipped peccary (*Tayassu pecari*), and paca (*Cuniculus paca*), to name just a few of the items on the Ache menu.

Cuategi fathered several children, became a grandfather, and experienced significant life changes, not the least of which was making first peaceful contact with the outside world when he was well into adulthood. Up until his twenties, he and his band lived solely within the confines of the Paraguayan forest, away from modern settlements, actively avoiding loggers, farmers, and poachers who sometimes ventured into the jungle—although from time to time he and his band were not above the occasional incursion into a farmer's manioc field when hunger made it a necessity. As he got older, his beard turned gray, he contracted and survived leishmaniasis resulting in the destruction of his upper palate, and lost more than a few teeth. Eventually he and other members of his small community decided to leave the forest to make first peaceful contact with a family of Norwegian missionaries, eventually settling into semipermanent housing on their land near the Ñacunday River in eastern Paraguay close to the Brazilian border. Today their life is taken up by the daily chores of farming, some modest foraging, and spending time with family and friends. In his seventh decade of life, Cuategi and his band now cope with the daily challenges of adjusting to a world full of cell phones, the Internet, bills, and diabetes. The rapidity and intensity of their transition from forest-dwelling, full-time foragers to blue-jeans-wearing surfers of the Web is dizzying and unprecedented in human evolution. Most are now on Facebook.

On the other side of the world in eastern Africa, a male chimpanzee makes his way through the Ugandan forest. He is large, robust, and emboldened by the life experience that comes from surviving four decades in the wild, avoiding poachers' snares and other potential threats. Although in relatively good health for his age, he is clearly more feeble compared to other males in their teens and twenties. The fur on his chin is white. He knows the importance of alliances and staying clear of neighboring chimpanzee groups that would readily beat him to death if he were foolish enough to be caught alone. During his twenties, he managed to hold the alpha position in his group and fathered several offspring but never devoted any time to caring for them. His life was preoccupied with staying out of the crosshairs of rival males and nurturing coalitions with other males who allowed him to outlive his rivals. As he becomes old and weak, he will disappear into the forest, never to be seen again. A primatologist may find his bones.

The lives of my father, Cuategi, and the chimpanzee could not be more different. Yet beyond their particular life conditions and environmental circumstances, the biological and social forces that shaped their journey of aging are similar. Three important factors unite them and how they aged. First, they are males. Belonging to sexually reproducing species, their lives were significantly influenced since conception by the union of their parents' X and Y chromosomes as well as the subsequent cascade of hormonal and other developmental changes that come with being a male. This is not to say that their genes determined their destinies or daily life decisions, but their sex chromosomes and male phenotype certainly had significant roles in their lifetime development. Second, the biology of their aging was shaped by evolution by natural selection. As a product of evolution, their aging processes share important similarities as a result of their common ancestry. Yet humans and chimpanzees are different because of the environmental and social challenges that face each species. As a species, human males also face common and distinct environmental and social challenges that shape their behavioral and reproductive strategies as they age. Finally, male aging is unique compared to that of women because of the different constraints that emerge from our reproductive and metabolic biology.

In this book, based on those three premises, I will argue that an evolutionary lens is vital for understanding the biology of male aging. Moreover, I will contend that evolution has shaped male health, influenced human evolution as a whole, and will guide where we are heading as a species. In essence, Darwinian fingerprints are on every aspect of male aging. But why should one care? There are numerous books on health and well-being as well as evolutionary biology. But understanding the evolution of male aging goes beyond health and well-being. There are other important reasons to understand male aging from a Darwinian perspective.

Besides putting some modest polish on our scientific literacy, there are other ways in which understanding male aging can be informative and more than a bit interesting. If you bought this book, it is probably because you have some interest in the topic. You're an older man, the partner of an older man, or someone who has a general interest in the evolution of the human condition. Perhaps as an aging man, you are looking for some insights into why and how your body is changing. That's a perfectly reasonable motivation. If your partner or other loved one is an aging man, you may be wondering why he is spending more time in his favorite chair, snoozing, or otherwise changing from when he was younger. An entire chapter is devoted to changes in muscle mass, the dreaded love handles, and the hormones that are responsible. You may also be interested in more serious aspects of health like prostate disease, erectile dysfunction, or male pattern baldness. Is testosterone supplementation safe? What does it mean to have low testosterone levels? What are "normal" testosterone levels for older men? These are all rational incentives to read this book.

But to gain a *deeper* understanding of health and illness, one needs evolutionary theory. The reason is that natural selection does not create perfect organisms, and imperfect organisms tend to have a nasty habit of accumulating defects, aging, and eventually dying. Men and women are also subject to different selection pressures that contribute to sex differences in aging and life span. That is, during our evolutionary past and arguably up to the present, men and women have faced challenges that are unique to their sex. Drawing on examples from reproduction, for women, these include childbirth and lactation. For men, it has been competition with other males and evolving ways to be attractive mates. Evolving optimal traits that can cope with the effects of aging is in itself a major challenge.

Virtually every physical trait is a compromise in response to the development and needs of other traits. Sometimes traits complement each other and actually promote the effectiveness of the function of other traits. Often the investment or expression of one trait will compromise the function of another. These compromises or "trade-offs" lead to imperfections that result in physical degradation, illness, and eventually death. Trade-offs are a primary driver of aging. Men and women contend with different trade-offs and therefore exhibit different patterns of aging and death. A male body needs to make crucial decisions about how to allocate calories and other resources to promote reproductive success. In Darwinian evolutionary theory, surviving is not enough. Survivorship is only time in service of trying to produce offspring and push one's genes through to the next generation. As we will see, men pay a significant tax in the form of shortened life spans in order to include their genes in subsequent generations. What is interesting about men, however, is that they may have developed unique solutions to maintain the ability to father offspring at older ages and address the challenge of somatic degradation and aging. More on that later.

Besides questions of health, well-being, and how evolutionary theory can shed light on these important issues, I would offer that there are other deeper and compelling reasons for taking an interest in aging males. If you are reading this, it is safe to assume that, like the author, you are human. Therefore, how evolution shaped our ancestors into the beings that they were and how that resulted in you reading these words is, in my humble opinion, bloody interesting. How did we emerge as the dominant species on the planet? Yes, there are far more beetles on the earth than people, but one cannot ignore how we have diverged from our other ape cousins and evolved traits that are unexpected for a large-bodied great ape. Simply by sheer numbers and our ability to shape our environment, for better or worse, we are pretty successful as a species. I will argue that to get to this point in our evolutionary history, traits associated with male aging may have been leveraged to facilitate the evolution of human-specific traits in all individuals, both men and women. These traits were vital to our success as a species.

Fast-forward to the present, and we can readily agree that the world is clearly run and controlled by men. I am certainly not condoning this reality; I am simply making a statement based on fact. Most of those men are also older, say, over the age of fifty. With very few exceptions, virtually all heads of state, chief executive officers, and individuals who wield socioeconomic and political power are older men. Clearly men have leveraged economic, social, and political power in their favor. How did this come to pass, what makes them tick, and how did they evolve? The acquisition of this power and influence is through competition with other males and active and passive subjugation of women's political and socioeconomic power. For anyone who is interested in gender equity or motivated to pursue it, I would suggest that understanding the evolution of older men and how they arrived at this point in history would be at best strategic and at least somewhat useful. But before we delve into these complex and nuanced questions, we will have to define what we are assessing in men.

WHAT IS AGING?

Understanding the specifics of aging in men requires that we get a handle on what we are putting under the evolutionary microscope. Aging is more than simply the passage of time or the number of candles on a birthday cake. Aging is a physical process that affects individual men and is guided by a number of factors including physics, genes, disease, and other environmental challenges. From a genetic perspective, the process of aging shares similarities with the biology of height. Both have a high degree of heritability, which means that the genetic complement of your parents has a significant predictive effect on your own height, or in this case how you age. The association is certainly not perfect, but the relationships are pretty strong with genetic variation accounting for 20–25 percent of the chance one will live past the age of eighty.[1] However, despite this high degree of heritability, no single gene or suite of genes accounts for this relationship, at least none of which we are aware. Genes are an important aspect of aging, but their expression often depends on environmental cues and the action of other genes. In addition, genes are carried around within individuals, populations, and species that interact with each other in crucial ways. We need a more holistic approach to understand the whole evolutionary picture.

Zooming out from the individual, we can also observe aging from a higher perspective, one that examines male life span and mortality both within and between populations and species. Men have longer life spans compared to male chimpanzees, for example; the probability of dying at various ages is remarkably similar. Comparative analysis of the demography between us and our closest primate relative reveals that there are deep evolutionary roots that guide male mortality, aging, and life span. In addition to our similarities to other primates and mammals, humans are also unique in ways that are clearly evident, such as how we walk, our lack of body hair compared to other primates and mammals, our large brains, and language, just to name a few spiffy characteristics. While these are all qualities that would make any self-respecting primate proud, there are other traits that allow scientists to assess human uniqueness on a more fundamental level. These characteristics are called "life history traits," which emerge from a branch of evolutionary theory called, not surprisingly, life history theory. In essence, this theory is an extension of evolutionary thought that provides scientists with a way of studying the evolution of different species by comparing basic traits that all organisms have in common. Of particular interest for the purposes of this book is the process of aging. All organisms deteriorate to some degree with time and are faced with the constraints created by this degeneration. While the entire body ages at once, different parts can age at different paces and in different ways. With life history theory we can ask: Are certain characteristics of aging in men unique to humans or are they common to other organisms? How has aging affected other life history traits in humans such as reproductive effort in both males and females? Addressing these questions also allows researchers to determine whether a trait is the result of some biological constraint in males or if it recently emerged in humans, perhaps as a result of changes in environment. These are deep, juicy questions.

Viewing aging as a life history trait is fairly straightforward. However, what gets lost in the shuffle is what one might consider an adaptation. In basic terms, an adaptation is a trait that is the result of natural selection. In humans, a large brain is all but certainly an adaptation since the advantages must outweigh hefty metabolic costs associated with having so much neural biomass between your ears. Another trait is our ability to walk habitually on two legs. Bipedalism is a unique and fancy way of getting around if you are a great ape. However, other organisms such as birds are bipedal, which kind of throws a wrench into the uniqueness of bipedalism. This raises a challenging issue that revolves around how we compare humans with other organisms. After all, we cannot say that humans are unique at anything if we do not compare a particular trait with that of other animals. What we need is a common currency of natural selection that can be compared between species regardless of evolutionary heritage, environment, or genetic complement. This is where life history theory provides a tremendously useful and powerful way of deploying comparative analyses of traits that are common to all life. The following life history traits are a common evolutionary currency that can be applied to virtually any comparative investigation. Besides aging and life span, they include distinct events and traits that are shared by all organisms. These include:

- Size at birth (or hatching)
- Rate and pattern of growth
- Age at reproductive maturation
- Adult body size
- Size and number of offspring
- Rate of reproduction
- Sex ratio of offspring (for sexually reproducing organisms)[2]

The power of life history has only recently been deployed in the service of understanding human evolution. However, since that time, tremendous strides have been made in addressing the evolution of these traits using observational, biological, and demographic assessments of various populations. Another core idea underlying life history theory is that all organisms, including humans, are faced with universal challenges that constrain and influence the evolution of life history traits. The central challenges are the availability and allocation of time and energy. As men age, time and energy availability becomes ever more constrained and challenging to manage. The idea of energy availability being a constraint emerges directly from one of the basic laws of thermodynamics; that is, energy spent in service of one purpose is made unavailable for other purposes. Time and energy seem to be in chronically short supply even today, although the emergence and spread of obesity is definitely a condition that has emerged from our ability to sequester too much energy in environments that are awash in quick, abundant, and easily accessed calories.

Understanding the harvesting and allocation of these resources within the physiology of an individual is what drives much of life history research. It is assumed that organisms are often limited in the availability of time and energy and must therefore make allocation decisions that directly or indirectly affect their fitness, through influences on either survivorship or reproduction. Life history traits, such as those related to aging, do not operate in a vacuum. Traits related to growth and reproduction often vary in response to each other as well as with the processes associated with aging. For example, the time and energy devoted to reproduction are hugely important. In life history parlance this is known as reproductive effort. As we will see, reproductive effort will have a tremendous say in how males age. Unfortunately sex does come with a price, not only for men but for all reproducing organisms.

Constraints on time are best illustrated by trying to be in two places at once. Try it. Doesn't work. Unless you are Dr. Who, you will fail miserably. Also try and do two things at once. Sometimes we succeed, although our distribution of attention often has mixed results. Most of the time we fail in spectacular fashion. It is almost impossible to whip up a batch of marinara sauce and simultaneously edit a manuscript, or take aim at a juicy monkey with your bow and arrow while conducting a debate on tribal politics. Doing two things at once requires skill, planning, and choosing the appropriate tasks to tackle together. Yet we are faced with these challenges every day. How we manage these decisions and live with the outcomes is important to our evolution.

In terms of energy, even if one is obese and literally swimming in coagulated calories in the form of fat, you cannot burn the same calorie for two different purposes. The same calorie cannot be metabolized to heal a wound while at the same time be used to replace a worn-out kidney cell. If you are fortunate, you may have enough energy to deploy two different calories to deal with these challenges. With finite and constrained resources of time and energy, every organism faces trade-offs. If you invest time or energy in one purpose, chances are you have made that time/energy resource unavailable for anything else. These trade-offs eventually manifest themselves in larger physiological negotiations that affect a greater proportion of the organism often in the form of aging. Selection favors organisms that reproduce effectively at higher rates. But to do so, you need time and energy. Those that allocate time and energy most effectively and efficiently will most likely have an advantage over other organisms and be favored by natural selection.

Life history theory can also provide useful perspectives on the various health issues men face as they get older such as prostate cancer, loss of muscle mass, and difficulties with weight management. The deployment of life history theory and evolutionary biology in general to gain a greater understanding of health and disease is known as evolutionary medicine. While this field has been instrumental in developing a greater understanding of various diseases, very little attention has focused on male aging. For example, in a later chapter we will discuss how reproductive biology early in life may influence the risk of acquiring age-related illnesses such as prostate cancer. Finally, we will turn the tables and explore how the evolution of traits in older men has affected the evolution of our entire species and how this may continue going forward as we continue to evolve.

HUMAN DIVERSITY

We need to appreciate the importance of human diversity. Most biomedical research focuses on urban populations who live more or less Westernized lifestyles, which usually means people who ingest lots of calories, are relatively sedentary compared to many other populations around the world, and live in high-density areas. Harvard evolutionary psychologist Joseph Henrich has dubbed these people Western, Educated, Industrialized, Rich, and Democratic, or WEIRD.[3] Information from populations who are not WEIRD and have different lifestyles is becoming more available, but the vast majority of our information on human biology comes from American and Western European populations. This research has been invaluable, but what has become apparent over the past few years is the broad range of physiological variation that is a hallmark of our species. My father and Cuategi led very different lives, yet if one were to compare the demographic patterns of aging and mortality of their respective communities, one would find striking similarities and important differences that influenced their aging. For example, as my father got older, he developed diabetes. Cuategi did not. Was it due to genetic differences? Perhaps, but their different lifestyles, activities, and diets likely had a greater impact on their risk of acquiring metabolic disease with age. To understand human evolutionary biology, including male aging, an anthropologist's perspective that looks at older men from across a broad range of the human condition is the only way to appreciate the adaptability of humans as well as the commonalities that bind us as a species.

In my present administrative position at Yale as deputy provost, I am responsible for helping schools and departments find excellent faculty in segments of society that historically have been underrepresented at Yale and in academia in general. Our goal as a university is to promote greater diversity since homogeneity of thought, perspective, or background is seldom conducive to scholarly growth and excellence. Diversity matters and is important. From a biological perspective, this is equally true. To understand male aging from an evolutionary perspective, we need to look at a very broad biological and ecological landscape. We need to look beyond our own shores and closely examine the evolutionary biology of older men across the spectrum of human existence. Anthropologists like me are particularly interested in hunter-gatherer communities and others who do not live as sedentary, well-fed individuals since their lifestyles are more indicative of the challenges humans have faced throughout our evolution.[4] This is not to say that hunter-gatherers are perfect models of the conditions during human evolution. Hunter-gatherer populations themselves exhibit a broad range of lifestyle and ecological variability from the Hadza, who live in dry scrub areas of Tanzania, to the Shuar, who inhabit the Amazonian region of southeastern Ecuador. Nonetheless, they do provide a useful lens through which we can get a glimpse of the conditions that were important during human evolution.

It is also vital to understand that biology responds to environmental variation. Compared to other great apes, humans are extremely diverse and adaptable organisms. We maintain an extraordinary ability to adjust our physiology, behavioral strategies, and social arrangements to accommodate environmental change. One only needs to note that humans have colonized most corners of the globe since leaving Africa about two hundred thousand years ago. In the case of earlier species of Homo, this wanderlust occurred even earlier. In contrast, our other great ape relatives have remained in fairly specialized environments. The potential for adaptability to the constraints of aging is no exception.

EVOLVING OPTIONS AND MALE AGING

So far we have used the word "trait" several times. Within evolutionary biology, traits that are relevant to male aging often refer to things like amount of muscle, amount of fat, and visual acuity—pretty much anything that you notice changing as one gets older. Strictly speaking we restrict our definition of a trait to characteristics that have the potential to evolve over time as a result of natural selection. Other important traits that are unique to older men that may have contributed to the evolution of our species include paternal investment, devaluing the importance of physical strength in men, and leveraging experience. Compared to other primates and great apes, these are traits that are unique to older men. As a consequence, older men are able to contribute to reproduction beyond the ability of fertilizing an ova. I will argue that devaluing the importance of physical strength allowed for longer life spans, inhibited the aging process, and decreased mortality resulting from environmental risks. As the importance of physical strength declined, knowledge and experience took on more central roles in the daily lives of older men. This will be an important topic in this book, one that we will revisit later on.

The contributions of older males to the evolution of human life history traits that define us as a species are also very noteworthy. Humans are strange creatures who failed to get the memo outlining the characteristics that are expected from any respectable creature who would call itself a great ape.[5] Using diagnostic traits such as body mass, food sources, and environmental hazards, an evolutionary biologist can wield some predictive power about what sorts of characteristics should evolve in a particular species. For example, orangutans are large bodied compared to many other mammals and primates. They also live in an environment where food resources are often unreliable and characterized by dramatic swings in availability. Consequently orangutans reproduce very slowly and have much slower metabolisms than one would expect given their body size.[6] When the food supply is unreliable and there are very few hazards in the form of predators, it makes sense to slow down and take it easy. And when there are few predators, as in the case of orangutans, a species can grow for a longer period of time and achieve a larger body size, which has many benefits.[7] In contrast, despite our large body size, humans have evolved very different life history traits.

Consider that there are over seven billion humans on the planet and fewer than a million other great apes. As a species, we are very efficient at reproducing and have done pretty well compared to other great apes. In addition, the human life span is much longer than one would predict. Life span is usually correlated with female reproductive life span; that is, when females cease reproducing, it is usually a signpost that mortality is imminent. However, in humans about a third of female life is post-reproductive. This is unique and begs the question: How did this evolve and have older men been part of this evolutionary development? I would argue that the answer is "yes."

There has been a common assumption that men tend to stop reproducing around the same time as women. This emerges from a somewhat ethnocentric perspective that is based on demographic data from Westernized populations such as the United States and Europe, one that fails to incorporate human cultural diversity. However, we will see that the American and European pattern of male fertility is not universal. Men maintain the capacity to reproduce long after similar-aged women have gone through menopause. Male ability to reproduce at older ages allows natural selection to operate and shape human evolution in ways that are unique compared to how other primates evolved. The question then is, if aging causes men to physically deteriorate, and physical condition is often important to reproductive success in many mammalian and primate males, how did men evolve the ability to reproduce at older ages? In fact, how did the human life span evolve to become longer than the reproductive life span? What are the implications to human evolution? This is a big deal.

LIMITATIONS OF EVOLUTIONARY THEORY

Before we get too excited we should tap the brakes a bit. Evolutionary theory and life history theory are tremendously useful, but there are limits. It is certainly not uncommon to have different schools of thought in any academic field. Try being in an elevator with an applied and theoretical physicist. Making it gracefully to the fifth floor is doable. Getting to the twelfth without a disagreement is an accomplishment.[8] The rift between scholars who deploy evolutionary biology to human research questions and those who critique this perspective is also quite animated and I believe based on a fundamental misunderstanding. This unfortunate division lies between the more subjective realm of academia and those scholars who are more scientific or biological. On more than one occasion I have been faced with a critic who takes exception to attempts to deploy evolutionary theory to understanding human behavior. Curiously there is very little pushback when it is used for non-human primates or to grapple with the fossil record. However, if there is any mention of evolutionary biology to understand contemporary humans today, more than a few eyebrows are raised and one is thankful to be nowhere near an elevator.

The unfortunate misunderstanding stems from the belief that evolutionary biologists are out to prove that everything is in our genes and that biology will explain everything. If we know the level of hormone x, we can explain complex behaviors like aggression, love, and our affinity for cats on YouTube. This is simply misguided. Contemporary evolutionary biology is firmly grounded in gene-environment interactions, and unless a change over time in a trait can be traced back to the necessary conditions of natural selection, there is not much room for evolutionary theory in the conversation. Genes, and therefore our biology, are often inextricably influenced by environmental factors such as diet, activity, social interaction, care, attentiveness, and any number of things we can think of that shape our lives every day, including culture. As I used to state in my undergraduate courses every year, when it comes to nature or nurture, the answer is "yes." On the flip side, believing that our brain, genes, or the various glands that secrete hormones do not influence our behavior is similar to believing our choice of clothing is not influenced by the weather. I may ponder whether to don my gray coat or my brown one, but I will surely put on a coat if it is a cold Connecticut winter day. We are products of the environment our ancestors experienced in the past as well as the conditions we face today. Those who take issue with evolutionary approaches to behavior are often led to these perspectives by the publicity of poor science. We won't do that here.

Evolutionary perspectives cannot address individual behavior and should certainly not be used as an excuse for bad behavior. This is particularly worrisome when the discussion revolves around men. One cannot point to the actions of an individual and say, "Oh, Darwin made him do it." Everyday life is just too complicated to make overly simplistic assertions. As someone who has spent a good portion of his career studying the evolutionary biology of men, few statements prompt me to duck into my local pub out of frustration quicker than "Boys will be boys" or "Oh, that's just testosterone poisoning." Evolutionary descriptions of human behavior and biology are not justifications. They are attempts to explain and understand. Indeed it is sad that Darwinian theory has been corrupted and twisted for political, sexist, and racist gain over the past century. Charles would not approve.[9]

As with most, if not all, scientific fields, biological anthropology has historically been a male-dominated discipline that has had a checkered history when it comes to research that is tainted or motivated by racism, classism, or sexism. An overabundance of caution is therefore merited when evoking an evolutionary perspective to understand men.[10] It does not mean it cannot be done; it just has to be done right. In this book, assertions to evolutionary biology will be made carefully and conservatively. Fortunately, the trait that we will be discussing, aging, is central to the evolution of all organisms since the biology that governs how we age has been at the core of natural selection. However, I will take some chances in the hope that the reader can make informed conclusions based on scientific evidence and critical thought. I am certain that some of my ideas will ultimately prove to be wrong. But I will assert that all merit consideration.

Obviously the biological aspects of human male aging are significant. But this does not mean that other aspects of the human condition have nothing to do with male aging. Humans are cultural beings. Some have argued that other organisms such as chimpanzees exhibit social behaviors that are similar to those found in human culture.[11] This may be true, and the arguments for comparative cultures are beyond the scope of this book. However, it is certainly true that no other organism exhibits the level of cultural complexity that is seen in humans. Humans interact with each other, develop unique and often illogical social practices, and do things that seem to contradict the premises laid out by Darwin and later evolutionary biologists. It may be tempting to dismiss cultural practices as simply noise that needs to be discarded in order to get at the real facts underlying male aging or any other aspect of human evolutionary biology. However, this would be a mistake. Culture and social behaviors are extensions of our phenotype, and while the connections with genes and phenotypic expression are complex and sometimes unwieldly, they are part and parcel of who we are as a species.

As we delve into the science of male aging and mortality, it is important to note that there are shades that exist outside the intellectual visibility spectrum laid out by evolutionary biology. The cultural and anthropological subtleties of male death inform our understanding of male aging and likely influence how men age and die. Clearly humans are more complex compared to most organisms, but the roots of hesitation to bring humans into the conversation about future evolution run deeper than ordinary scholarly banter. Reluctance to apply evolutionary theory to questions of human biology stems from an anxiety of history. Anthropologists in particular are still licking their wounds from a legacy of poor science that was steeped in racist, sexist, and classist agendas.[12] Not so long ago, some ill-mannered scientists with social and political agendas peed in the pool of human evolutionary biology research. Some probably had no particular social or political agenda but were simply products of their time period. Nonetheless, since those missteps, many anthropologists are not eager to jump back into the water. I propose the filter of time has run its course and it is safe to wade in a bit, albeit with caution and vigilance for the ill-mannered types.

Now that we are duly primed on some basic background information, we are ready to discuss male mortality and aging. Onward.

CHAPTER 2

DEAD MAN'S CURVE

> It is possible that death may be the consequence of two generally co-existing causes; the one, chance, without previous disposition to death or deterioration; the other, a deterioration, or an increased inability to withstand destruction.
>
> —Benjamin Gompertz, *Philosophical Transactions of the Royal Society London*

Measuring aging and death seems like a grim task. But we have to embrace this necessity if we are to make any sense of male aging and mortality. At least that is what I used to tell my undergraduate students in my human evolutionary biology class. As part of their course requirements, they had to visit Grove Street Cemetery, which is all but embedded within the Yale University campus. The cemetery is the resting place of many who lived in New Haven and is actually a pleasant place to visit on a nice autumn day. I dispatched my students to the cemetery to conduct a simple demographic exercise on how to assess mortality patterns by recording the sex and dates of birth and death from the gravestones. This exercise, if done carefully, always reveals two things. First, life was pretty tough in the early colonial and industrial age periods. There are many babies and young children interred there. Second, men often died at an earlier age than women did.

If we had a larger sample size and more detailed information on causes of death, we might be able to observe other important differences in mortality between males and females in Grove Street Cemetery. These differences underlie some of the important contrasts between men and women and how they age. In this chapter we will build on the basic methods deployed by my students at Grove and try to get a bit more resolution on the issue of male mortality, including how it is assessed and how it changes over a lifetime. But before we delve into the juicy bits about how male life spans and aging differ from those of women, we need to get a handle on how humans age in general, how we measure aging, and how our aging is different from that of other organisms. These questions are vital to our discussion since they reflect the unique perspective that evolutionary biology can contribute to not only the larger issue of male aging but debates surrounding many other important health issues as well.

It may help to think of life span as the fuel gauge in your car. Many things such as road hazards, traffic, imperfections in the car's engine, detours, and weather conditions will influence how you expend that fuel and how you plan your journey. But there is one catch. You cannot refuel. Gas stations are nowhere to be found. This single but important detail will influence your decision making, risk assessment, and tolerance, as well as your choice of destination. As a relatively healthy male in my early fifties, my fuel gauge is past the halfway mark, assuming I do not encounter some fatal traffic jam. Differences in the initial amount of fuel in the tank reflect variation in life span among sexes, individuals, and species. Women can take comfort in the fact that they tend to have more "fuel" than men and by that we mean a greater probability of living to see another year.

We also need to consider the fuel gauge. How do we measure aging and mortality? For our purposes we can consider two methods of doing so. One is to measure various fluids and molecules to determine how much an individual has aged and how close he is to meeting his ultimate demise. The second method is to take a more encompassing view of mortality across numerous individuals, in essence, measuring the probability of death at a given age. This is the method my students used for their assignment and is part of the rich field of demography. When the methods of physiological and population measurements of mortality are merged you get the cool field of biodemography. With these tools we can not only assess the strength of natural selection at various ages but also have a pretty good sense of the biological mechanisms that are driving these patterns, including differences between males and females. The title of this chapter provides a sense of what this chapter is about: demographic curves that describe aging and mortality.

However, analogies only get us so far. If aging is a terminal condition, why hasn't natural selection allowed us to live forever? The old adage about death and taxes is only somewhat true. While natural selection cannot keep the tax collector at bay, it can create organisms that seem to tolerate aging better than others. To understand male aging, we need to have a sense of the range of options that have evolved in nature. Some organisms age very slowly, while others do not seem to age at all. How can we make sense of this? For that, let's head to the local drive-in theater.[1]

ZOMBIE CELLS

My parents used to pack the family in the car and head to the drive-in theater on warm summer nights. It was a great treat. On one foray, when I was four years old, we watched *Night of the Living Dead*, the original zombie movie made in 1968. I don't remember much except the smell of the vinyl backseat of my parents' Dodge Dart as I tried, unsuccessfully, to burrow my way into the trunk to avoid the sights and sounds of the dead feasting noisily on the living. Despite this initial aborted attempt at being exposed to the living dead, I became a chronic fan of the zombie genre of film and television long before it became vogue.

More recently, in the popular television show *The Walking Dead*, a plague has enveloped the world (or at least the state of Georgia, where the series is filmed), causing the dead to reanimate and wreak havoc among the living, usually in the form of attacking, biting, and consuming the latter. As the main characters struggle from one crisis to another in this post-apocalyptic world, they discover that one does not need to be bitten to become the undead. Every living person is infected by this unknown zombifying agent, and even if someone dies naturally, he or she will reanimate and join the ambulatory undead. In essence, all of the live ambulatory characters are dying, hence the title of the show.

While no zombie virus has (yet) infected us all, we are all "walking dead" since we are all dying with the mere passage of time through the process of aging. We dodge various "bullets" every moment of our life. But even in the absence of environmental hazards, our risk of death at any given time is greater than zero since a physiological imperfection can lead to death at pretty much any time. Sorry, I did say there would be some grim moments. Aging is a chronic condition that is managed through life, not unlike many other conditions that increase mortality and are considered "diseases." If we consider aging to be a disease, why not cure it? We can ask, why does death by aging exist at all?

At face value this may appear to be a silly question since most of us assume that aging and death are among the most inevitable processes in nature. However, silly questions sometimes yield intriguing discussions resulting in the revelation that the questions are not so absurd after all. While getting older with the passage of time (relative to the velocity of the observer and observed) is a function of the physics of the known universe,[2] degradation in association with age can and does vary from rapid aging to virtual immortality in a variety of organisms. Immortality can theoretically be a viable evolutionary option.

In the 1950s, Nobel Prize–winning evolutionary biologist Peter Medawar challenged traditional biological thought by asking why immortality was so rare. Armed with the knowledge of the New Synthesis, which melded evolutionary theory with the burgeoning field of genetics,[3] he postulated that natural selection should be expected to favor organisms that have more efficient repair mechanisms and therefore longer lives. This assumption suggests that in theory, there is no reason to automatically assume that all organisms should ultimately succumb to aging. Under the right conditions, selection should result in organisms that live forever. The question is, why don't these mythical creatures exist?[4]

An interesting aspect of aging research is that to study organisms in a coherent manner, an observer needs to be able to conduct comparative observations between individuals. In other words, an observer needs to be able to distinguish between different organisms. While distinguishing between individual persons is straightforward to you and me, it is not so easy with other organisms, in particular, unicellular organisms such as bacteria. When bacteria reproduce, they simply double their genetic material and split, creating two daughter cells through the process of mitosis. There are no mating rituals, pieces of jewelry to exchange, or bottles of wine to uncork. It is all quite elegant and efficient. It is, however, very difficult to distinguish between the progenitor or "original" cell and the daughter cell since both are virtual clones of each other. It is certainly possible for unicellular organisms to die off, perhaps as a result of the accumulation of unrepaired genetic damage that accrues with each division, but aging in unicellular organisms is quite different compared to that in sexually reproducing species like humans, who generate and reorganize genetic variation with every reproductive event.

Diminished aging in unicellular organisms is similar to what is found in immortal cell lines, that is, somatic cells that divide mitotically in humans and other organisms.[5] Under normal circumstances, mitosis of these cells is kept in check by many processes such as apoptosis, or programmed cell death. Before a cell can run amok and reproduce uncontrollably, our bodies engage genes that push a self-destruct button. Disruption of this process is what commonly leads to uncontrolled cellular replication, or cancer. The wonderful biological irony is that cellular death can and does keep us alive.

This is an aspect of research biology that received notoriety recently with the case of Henrietta Lacks, an African American woman who died from cervical cancer in 1951 but whose HeLa cancer cells have been maintained to this day through careful laboratory cultivation as an immortal cell line for cancer research.[6] Setting aside the important ethical aspects of this story, the HeLa cell lines illustrate how, under the right conditions, cells can be maintained almost indefinitely in the absence of some catastrophic event like a clumsy laboratory technician or rogue spritz of bleach. Unexpected environmental hazards such as these foreshadow what we will later distinguish as intrinsic and extrinsic mortality.

For now we will set aside unicellular organisms or cell lines and retrain our attention to aging in sexually reproducing species like ourselves. The potential for asexually reproducing organisms to achieve immortality illustrates the important role of sexual reproduction in our aging journey. Our aging story is going to be intertwined with sex. That is, aging is particularly germane to sexually reproducing organisms. In addition, sexual reproduction and overall energetic investment in reproductive effort will affect the rate of aging as well as mortality in general. We will therefore bid adieu to our unicellular friends and restrict our discussion to sexually reproducing organisms.

SEX, TRADE-OFFS, AND DISPOSABLE BODIES

Thanks to mom, dad, evolution, and sexual reproduction, the aging clock begins to tick once we are conceived. Since a significant proportion of the conceptus biomass comes from the mother's ova, which have been in biochemical stasis in her ovary since she was in *her* mother's womb, one can argue that the aging process was well under way even before your father's sperm tipped his biochemical hat to your mom's egg. As time passes, who and what we are inevitably change with each moment regardless of the quality of the surrounding environment. No matter how pristine or hazardous, the probability of surviving another second, minute, day, or year is never certain. Place an individual into a perfect environment without hazards and with the perfect diet, cognitive stimulation, and every resource that one could identify to maximize life span and you will still eventually end up with a corpse.

Before we go any further, we should define "aging" and "senescence." Simply put, senescence is the passage of time that can be measured by calendars, stop watches, or birthday candles.[7] Aging is the decline in physiological function that occurs with the passage of time. Peter Medawar used the analogy of the Oscar Wilde literary character Dorian Gray to illustrate this point. Though Gray grew older with the passage of time, only his painting exhibited outward signs of aging.[8] Are there general theories of aging? Yes, and although none is perfect, each provides a different but useful lens into aging. Especially useful are the exceptions to general principles of aging that give evolutionary biologists important clues about where to look next.

Transitioning from the cellular level to whole organisms and species, the biology of aging is rooted in some basic observations. First, small mammals tend to have accelerated aging and shorter lives compared to larger mammals. Think of the brief and frenetic life of a mouse, the longer, more laid-back life span of an elephant, and everything in between. Why the variation? The commonly accepted answer is that small organisms maintain faster metabolic rates to keep up with heat loss in relation to their surface area and basically burn out faster. That is, the amount of tissue, cellular, and genetic damage that occurs as a result of the rapid processing and turnover of energy outpaces an organism's capacity to repair itself. Larger mammals are able to conserve heat more efficiently, have slower metabolic rates, burn energy slower than smaller animals, and live longer. This is commonly known as the rate of living theory.[9] An important parallel is the difference in metabolic rate between men and women in association with life span and aging. Despite being larger on average compared to women, men tend to burn more energy per unit time. Does this influence sex differences in life span? Perhaps.

While the rate of living theory tends to hold true for most mammals, there are important and very curious exceptions that merit attention. Among mammals, bats, naked mole rats, and lemurs did not get the memo. They all live much longer than expected for their body size.[10] They also live in environments that have fewer hazards than those of most other mammals. Keep this in mind. Similarly, many birds are evolutionary nonconformists, smugly thumbing (?) their beaks at the rate of living theory. They have fast metabolic rates yet some live a very long time. Consider parrots, which can live for decades and sometimes outlive their owners. Clearly the story of aging involves more than body size and metabolic rates.

Not long after the Soviet Union launched *Sputnik*, another idea was put into evolutionary orbit. Eminent biologist G. C. Williams expanded on the evolutionary biology premise of pleiotropy, which is the idea that physical traits are not independent of one another. If you alter one trait in some way, perhaps through natural selection, that change has an effect on some other trait. In other words, everything is connected. Think of it as evolutionary side effects. He went further, noting that biological processes that have beneficial effects on fitness early in life may have detrimental effects later when their influence on reproductive success wanes with age.[11] In other words, at older ages when reproduction is on the decline or ceases altogether, there is no reason to invest resources toward staying alive if there are no fitness benefits. Williams postulated that this is the reason some traits that support reproduction earlier in life may cause serious health problems that shorten life spans later on. He called this antagonistic pleiotropy. An example would be high ovarian steroid levels in women that are necessary for fertility. Yet a lifetime of exposure to these hormones may increase the risk of breast cancer at older ages.[12] Similarly, a lifetime of high testosterone exposure might augment a man's ability to reproduce during early adulthood and then accelerate aging and increase the risk of prostate disease later on in life. We will revisit this in the chapter on health.

Around the time my wife and I were first pulling into Cambridge, Massachusetts, in our blue Toyota pickup truck after a ten-day cross-country journey to start graduate school, evolutionary biologist Thomas Kirkwood published a theory that expanded on the antagonistic pleiotropy contribution of Williams by suggesting that, since bodies are disposable, they should only last as long as what would be optimal in relation to fitness benefits. If individuals can optimize their fitness by living five years instead of ten, then that life span would have a selective advantage despite being shorter. What would be those advantages? Kirkwood made the astute connection that all organisms need to allocate energetic resources among various life history needs such as growth, maintenance, and reproduction. If there are advantages to investing these finite resources toward reproductive fitness over maintenance, then organisms should do so even if it compromises life span and accelerates somatic deterioration due to aging. This is called the disposable soma theory.[13] This idea is especially germane for men since they commonly put their bodies at risk if there is a reasonable possibility of gaining access to mating opportunities and increasing fitness. To put it bluntly, men are willing to bet a stack of survivorship chips if the big payout is the possibility of sex.

Besides the immediate possibility of sex, what other factors would cause a male to accelerate and increase investment in reproduction to the detriment of aging and longevity? If an individual finds himself in a hazardous environment, it may well be wise to invest in faster and earlier reproduction since the risk of death threatens to cut life short. In other words, do not put off for tomorrow what you can do today. However, there are distinct advantages to reproducing earlier and faster regardless of the condition of the environment. If organisms have this capacity, why don't they do it? The answer is that there is a serious cost to reproducing in the form of mortality and accelerated aging.

Several laboratory studies have successfully tested the prediction that reproduction accelerates aging and increases mortality; however, I am prone to pay a bit more attention to what is observed in the natural world. A classic field study that tested the disposable soma theory focused on the humble opossum. These aren't exotic animals like naked mole rats or lemurs, which may be examples of unique evolutionary solutions to the challenge of aging. Instead, opossums are common critters that live in a variety of environments that lend themselves to natural comparative experiments that can shed light on the question of the influence of environmental hazards on reproduction. Evolutionary biologist Steven Austad studied two populations, one that lived on an island with no predators and another genetically similar one that lived on the mainland and was subjected to significant predation by coyotes and such. In short, Austad found that the low predation (extrinsic mortality) island opossums reproduced later and exhibited physiological signs of slowed aging. The mainland opossums, on the other hand, started reproducing sooner and more intensely, and exhibited signs of accelerated aging.[14] Making opossum babies earlier and faster is costly.

Are the detrimental effects of reproduction on aging also evident in humans? Perhaps. I witnessed many young Ache girls age rapidly once they started having children. The daily grind of activities necessary to care for a family surely contributes to their physical decline. Public health researcher Arline Geronimus dubbed the process of declining health in response to early life challenges as "weathering," which is a very appropriate analogous description. Indeed I once gave a lunchtime lecture to a group of Yale employees about the possibility that having children causes mothers to age faster. The many moms in the room responded with hearty laughter and a rolling of the eyes to the silly professor. Clearly this was no surprise to them.[15] Scientists, however, need evidence to support what might seem obvious to many.

My research group at Yale University recently completed a study that hypothesized that women who have more children will exhibit physiological signs of accelerated aging. Demographic evidence supports the hypothesis that having more children leads to shorter life spans in women, but the contribution of physiological factors is not well understood.[16] To test this hypothesis we assessed oxidative stress, a key physiological biomarker of tissue, cellular, and genetic damage that contributes to aging in all organisms in association with the lifetime number of children. Our participants consisted of rural postmenopausal Polish women who are part of a long-term investigation of women's health and evolutionary biodemography run by my collaborator Grazyna Jasienska of Jagiellonian University in Krakow, Poland. Because all of the women were postmenopausal, the hazards of childbirth were not a confounding factor. There was also no history of contraception since this is a devout Catholic population. We found that women with more children had significantly higher levels of oxidative stress compared to those who had fewer children. To our knowledge this is the first physiological evidence of the predicted trade-off between reproductive effort and aging in humans.[17] Yes, having children contributes to aging in women. This is all very interesting, but what about the guys?

In contrast to women, men do not bear the metabolic or oxidative stress costs of pregnancy or lactation, and it has been commonly assumed that the costs of reproduction in men are miniscule with only a few calories being thrown into service for spermatogenesis. However, in the field of evolutionary biology, there is the propensity to expect equilibrium in biological systems. How can it be that women carry a disproportionate amount of the cost of reproduction in our species? This disparity would be unsustainable and evolutionarily unstable. Somehow men should be shouldering a proportionate amount of investment in reproductive effort.

Instead of gestation and lactation, men wear their reproductive investment in other ways, specifically sexually dimorphic muscle tissue. Consider the following question. Compared to women, why do men carry around 20 percent more active metabolic tissue? This is a significant metabolic cost especially when added up over a lifetime. The short answer is that male reproductive effort is reflected in tissue that is maintained to augment competitiveness, attractiveness, and the abilities necessary to procure important resources. Sexually dimorphic muscle is as metabolically expensive as pregnancy and lactation when measured over a lifetime. This tissue tends to wane with age, however. What is a man to do? Perhaps shift strategies? I will argue yes.

The social and ecological world of human males is fraught with greater risk and higher extrinsic mortality compared to that of females. As we will see, evidence of consistently higher male mortality in modern populations, in hunter-gatherer populations, in primates, and from the archaeological record all provide compelling evidence that higher male mortality due to environmental and behavioral hazards compared to female mortality has a significant negative impact on survivorship.[18] Many of these mortality and morbidity differences are self-inflicted, but the result is still the same. Which begs the question: Why would males put themselves in situations and/or conditions that would compromise survivorship, especially when female behavior demonstrates that there are alternatives to putting oneself at risk? In many cases, higher extrinsic mortality among men is likely to have contributed to their intrinsic mortality patterns; that is, consistently higher extrinsic mortality and shorter life spans compared to those of women result in physiological deficits that accelerate aging and shorten life span. How could this evolve?

Natural selection does not necessarily favor physiological processes that make us feel better or healthier. If a trait evolves that allows males to have more offspring or perhaps allows for those offspring to be more likely to survive and reproduce, it is quite likely that there will be some nasty surprise down the road. There is always a catch. Sex differences in the cost-benefit trade-offs among reproductive effort, extrinsic mortality, and potential fitness payoffs are significant. Since a primary constraint in mammalian males is access to mating opportunities, the potential payoffs for risky behavior or metabolic investment that eventually erodes survivorship are greater for males than for females.

In other words, a male who engages in a behavior or invests in metabolic processes that increase his present mortality risk by 1 percent but also promote his probability of mating opportunities will have potentially greater fitness benefits compared to a female who engages in the same risk and mating benefit since access to more than one mate will not increase a female's fitness. There may be other benefits for females such as additional provisioning or offspring care, but having additional offspring is not one of them.[19]

Now that we've laid down a sound theoretical foundation, we need tools to test the hypotheses that emerge from these ideas. Aging research is embedded in physiology and demography, that is, measuring the physiological effects of aging and assessing patterns of mortality in populations. Here we delve into the "curves" of mortality. Let's measure death.

MEASURING MORTALITY

Physiological biomarkers such as those that reflect oxidative stress provide us with a particular focal lens that shows what is happening within individuals. However, evolution involves changes at the populational level. Natural selection acts on the individual. Populations evolve. To understand the evolutionary aspects of aging and mortality, we need to measure these factors across populations and species. How do we do this? This seems like an odd question since the obvious answer might be to simply count the number of deaths in a population. This seems like a good place to start, but death in humans is more than demographic assessments of mortality. To fully appreciate and delve into the evolutionary anthropology of death, and how it fits into aging in men, a more nuanced approach would be more informative and, in my opinion, more interesting. First we need some basic skills to assess mortality and death. Then we will look at patterns of mortality and life span, specifically differences between men and women.

There are basically two ways to measure and quantify death. One can count the number of deaths in specific age classes. In very rough terms, for example, we can assess the risk of death in younger individuals compared to older individuals. Interestingly the mortality patterns mirror each other, albeit for different reasons. From this measure one can assess the probability of death and the likelihood of reaching a certain age. This can be envisioned as running a gauntlet with various hazards being faced at each stage. The hazards can be external or internal (the latter would result from physical imperfections).

Depending on the research question, biologists also are interested in the life span of an organism. That is, what is the common shelf life of an individual in a given species? Life span is fairly easy to assess. One only needs to record birth and death dates from the headstones in any cemetery to compare the life spans of men and women. It is not as easy to do this for other great apes. Life span can sometimes be garnered from captive populations, but getting accurate data from wild populations is much more difficult. Nonetheless, the efforts of several generations of primatologists have allowed us to address this question.

Anthropologists Michael Gurven and Hillard Kaplan have coalesced a remarkable amount of information on mortality patterns among various hunter-gatherer societies. In their analysis, they document that life as a forager is challenging to say the least compared to modern lifestyles, especially for children and women of childbearing age. Nonetheless, forager populations exhibit important similarities to those of WEIRD populations, including high mortality during infancy and after the age of fifty.[20] To garner a fuller, more comparative perspective, let's look at one forager group in association with our closest evolutionary cousins, chimpanzees. Anthropologist Kim Hill and his collaborators described mortality patterns across the life span of two species in different environments: human hunter-gatherers represented by the Ache of Paraguay and chimpanzees.[21] Why hunter-gatherers? The assumption is that the social, energetic, and environmental challenges that hunter-gatherers face during everyday

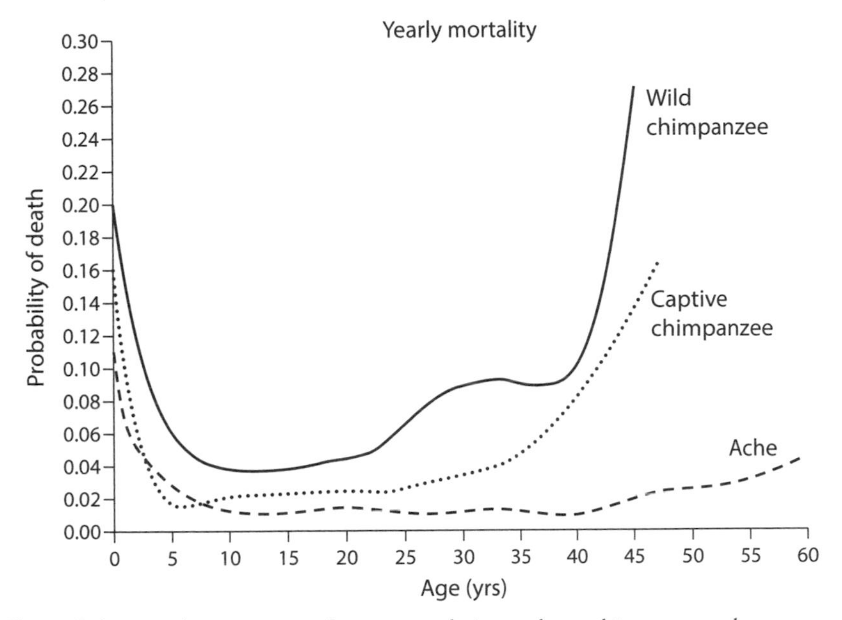

Figure 2.1. Mortality curves in a forager population and two chimpanzee cohorts. Curves are smoothed using a spline function.

Reprinted from K. Hill et al., "Mortality Rates among Wild Chimpanzees," *Journal of Human Evolution* 40 (2001): 437–50, copyright 2001, with permission from Elsevier.

life are more reflective of the conditions under which humans evolved over the past couple of million years compared to the modern conditions we commonly face today.

In these two great ape species (humans and chimps), we see the classic U-shaped mortality curve, demographic expressions of mortality, averaging both sexes, in contemporary hunter-gatherers, captive chimpanzees, and wild chimpanzees (Figure 2.1). In both humans and chimps, mortality is high early in life, attenuates with maturity and adulthood, and then increases dramatically at older ages. One can get the sense of this function by imagining a ball rolling down a U-shaped valley. In males, there is also a significant bump in mortality during late adolescence and young adulthood, something we will address a bit later. Interesting differences can be seen between humans and chimpanzees. Both wild and captive chimpanzees exhibit a sharp rise in mortality around the age of forty. By the time they approach fifty, the probability of death rises exponentially; most chimpanzees die before they reach their fifth decade of life.

Humans, however, go off the graph, exhibiting increasing but much lower rates of mortality after the age of fifty compared to chimpanzees. This sharp rise in mortality at older ages indicates that a species is approaching the end of its natural life span. Such demographic analyses illustrate an important point: all organisms tend to have species-specific expiration periods or semi-predictable finite life spans. For humans, regardless of environment, mortality starts to increase around the age of fifty but starts to spike after sixty-five. Does this mean that scientists can predict when you will die? No. But we can say that your chances of seeing your next birthday start to decrease after the fifth decade of life, with an expectation that this increase will accelerate with time.[22]

The importance of including Ache foragers is to illustrate that despite the hazards of living in a forest environment, the Ache still have life spans that are not that different from yours or mine. The increase in extrinsic mortality due to environmental hazards, snakes, accidents, and so forth results in the entire curve being higher. This is also evident in the contrast between wild and captive chimpanzee populations. The solid line representing the wild chimps is uniformly higher than the captive chimps' dotted line. This difference reflects the fact that chimpanzees in the wild face more hazards on a daily basis and thus have overall greater extrinsic mortality. Predators, disease, and the daily risks associated with getting food, mates, and all of the things necessary for life take a toll. This is not to say that being a captive animal is not without cost. Captive chimpanzees simply do not face the same mortal risks that are commonly encountered in the forests and savannahs of Africa. Yet their risk of death at older ages is very similar, suggesting that there is some shared aspect that affects their risk of mortality whether they live in the forests of Uganda or a New Zealand zoo.[23]

A more recent study of a smaller community of wild chimpanzees at the Kanyawara field site in eastern Uganda suggests that even between wild communities, mortality can differ dramatically. Anthropologists Martin Muller and Richard Wrangham report that mortality overall is much lower at Kanyawara compared to the four population aggregate samples included in the study by Hill and colleagues. However, the human comparative sample represented by Hadza hunter-gatherers of Tanzania also shows lower mortality compared to the Kanyawara chimpanzees, which is in alignment with the Ache results from the Hill study.

These human-chimp comparisons illustrate three important lessons. First, demographic assessments can provide important and very useful insights into comparative aspects of aging. That is, to determine if human aging is unique, we need to compare ourselves under different environmental conditions with other species. For the most part organisms exhibit the U-shaped pattern seen in Figure 2.1; all organisms, including chimps and humans, face similar biological constraints regardless of environmental differences or evolutionary ancestry. Second, these comparisons can aid in distinguishing between intrinsic and extrinsic sources of mortality. That is, we can readily differentiate between sources of mortality that are driven by environmental risk and those that emerge from physiological processes. Third, they lead us to pose other questions; for example, what biological mechanisms underlie the differences in aging that we see between species and between sexes? Putting these sources of mortality together, we can now assemble a more cohesive picture of how they all interact and how they contribute to male mortality and aging.

Extrinsic mortality can have a profound effect on intrinsic mortality since an increase in the former results in a greater probability of dying at any given time, resulting in a greater urgency to reproduce, which usually comes with risks and costs that can increase the likelihood of death through the acceleration of the aging process and exposure to environmental hazards. A classic example of this is seen in Trinidadian guppies. Evolutionary biologist David Reznick experimentally exposed wild guppies on the island of Trinidad to local common fish predators. Controlling for the availability of food and other potential confounds, Reznick found that guppies exposed to predators reproduced at earlier ages and accelerated their pace of reproduction.[24] The idea is that extrinsic mortality cues stimulate organisms to divert energy away from processes devoted to maintenance to reproduction.

Does this increase in reproductive effort in response to higher extrinsic mortality result in accelerated aging? Here the story becomes a bit mixed. Among these same guppies Reznick found limited evidence for accelerated aging.[25] Evidence for faster aging in guppies that reproduced sooner in response to predation risk was limited to the guppies that exhibited poorer swimming performance, which reflects poorer muscular function.[26] Clearly the relationship between investment in reproduction and aging is multifactorial and complex. For example, in the case of humans,

Grazyna Jasienska has suggested that variation in nutritional status between study populations may contribute to contradictory results in the literature on the role of greater reproductive investment on aging.[27]

For males, extrinsic mortality tends to be higher compared to females at particular life history stages. This difference emerges from both behavioral sources, primarily in the form of risky behavior, and biology, likely resulting from hormonal and other physiological differences that compromise male longevity. But why would such a disconcerting pattern evolve? Why don't men and women have the same life spans, and why do males tend to expose themselves to risky endeavors that compromise life span? Here we turn to why my wife will likely be choosing finger sandwiches for my funeral reception and not vice versa.

SHORTCHANGED?

It is quite likely that my wife will outlive me, and not because her diet or activity patterns are very different from mine. We tend to take our meals together, eat similar foods, and go for the occasional jog, we are about the same age, watch the same bad TV shows, and have similar lifestyles. Yes, I'm very lucky to be paired with my wife. Yet by the simple virtue of being male, I have less of a chance each day to see another sunrise; this will continue to be the case until the day she picks out my headstone. Behaviors, environmental hazards, and physiological factors all contribute to this disparity. It is important to note that sex differences in mortality often flip to the detriment of women during childbearing years, which provides additional insights into the role of reproduction in aging and mortality.

In the meantime, what is fascinating is the ubiquitous nature of the life span disparity regardless of culture, population, and even species. Despite the deep and broad literature on this topic in evolutionary biology, the causes of sex disparity in life span continue to vex health professionals. To understand shorter male life spans, evolutionary biologists need to examine patterns of mortality throughout a male's life. A life history perspective is needed, one that not only compares humans with other species but considers the effects of environmental hazards, reproduction, and other factors throughout an individual's lifetime.

Males tend to die at higher rates at younger ages than do women from infancy, throughout adulthood, and into old age, regardless of culture or environmental context.[28] It is fairly obvious that male mortality is often higher because males tend to do stupid things. One only has to watch Internet clips of snowboarders and inline skaters breaking bones and making frequent trips to the emergency room to see that surviving the teen years is a considerable feat for young males. Suffice it to say that they engage in behaviors that are not conducive to good health. Why?

Even if a male survives the gauntlet of skateboarding, the urge to purchase a motorcycle, and other unhealthy behaviors, he will continue to engage in behaviors that defy common sense. Men smoke and drink more than women, gamble away resources, and generally have bad health habits, including not going to seek care when they are ill. How does this happen? After all, humans evolved big brains. One would think (with this large brain) that when one gets sick, seeking care and treatment would be, well, a no-brainer. In a world where poverty is unfortunately too common, lack of resources and access to quality health care are important factors to consider. However, men fail to seek medical attention at the same rate that women do even when resources are not an issue. Yes, there are numerous exceptions, and it is important to remember that women face many health challenges that are specific to their sex, especially with regard to childbearing. But the general trend is that men simply do not take care of themselves as well as women do.

Assuming a man survives his teens, visits his physician regularly, does not smoke, refrains from drinking alcohol, and lives a relatively healthy life, he will still pay more for health insurance compared to women since he is more likely to become ill and die before women born at the same time. Continue to bear in mind that women shoulder sex-specific burdens that incur significant costs to health and well-being. Childbirth, autoimmune disorders, and male violence are all important causes of morbidity and mortality among women that certainly merit increased and continued attention. However, overall, men exhibit clear signs that they have a predisposition to devote time and energy to behavioral and physiological processes that inflict a clear cost on survival and neglect to invest time and energy in behaviors and physiological processes that would result in a few more birthday candles. In our society, which is obsessed with health care, well-being, and medicine, such a quandary has garnered the attention of the clinical research community. An article in the *New York Times* addressed these very issues, echoing the sentiments of most physicians: "'We've got men dying at higher rates of just about every disease, and we don't know why,' said Dr. Demetrius J. Porche, an associate dean at Louisiana State University's Health Sciences Center School of Nursing in New Orleans."[29]

Indeed, Dr. Porche is correct, although I would quibble with one word in his statement: "We." Unlike the vast majority of the medical community, evolutionary biologists have been quite aware of sex-biased mortality in numerous species, including humans, for quite some time. With very few exceptions,[30] clinicians have not considered the role of evolution much less life history trade-offs in their efforts to understand mortality and aging in men or disease in general. Moreover, the cause of this sex disparity in mortality throughout a male's life is interesting but not mysterious. In fairness to the medical community, the exact biological mechanisms for skewed male mortality are not completely understood, but comparative research of other organisms has provided evolutionary biologists with a compelling lineup of suspects. For example, differences in hormones, immune function, and metabolism are well-known culprits. Psychological experiments and brain scans also provide compelling evidence that men's brains operate in a way that often puts good sense and a healthy regard for well-being on the back burner.[31]

At a deeper level, the ultimate or evolutionary reasons are far from mysterious. Since male-biased mortality is common across cultures, human conditions, primates, and many other organisms, one can hypothesize with a fair degree of confidence that there are deep evolutionary underpinnings to male mortality that extend back millions of years when sexual reproduction first emerged. Taking this a step further, if there is a biological context to male mortality then it would follow that natural selection has had a hand in the emergence and persistence of male mortality and, by extension, aging. This may seem contrary to what Darwin would have predicted, but not really. Darwinian evolution is more about reproduction than survivorship. For these patterns of male life histories to evolve, there had to have been some evolutionary advantage for the physiology and neurobiology that underlie male patterns of death and dying.

In my first book, I titled a chapter "We're So Skewed," which discusses why sex ratios at birth are biased toward more males. This is a curious phenomenon since it raises several very interesting questions, many of which remain unanswered since my book was published over eight years ago. First, how does this happen? If we are to assume that there is an equal chance for an ovum to be fertilized by an X- or Y-bearing sperm, the ratio should be 1:1. However, it isn't, which means that there may be an unequal number of X- and Y-bearing sperm, the fertilization process is somehow biased, or there is unequal attrition during gestation, or all of the above. Suffice it to say that the answer remains elusive.

It is also widely assumed that equal numbers of male and female infants are born. Indeed, this would make sense since in sexually reproducing species, such as humans, one male for every female would be the optimal arrangement. However, such an equal arrangement is largely absent. At birth, a slight but consistent bias toward more males is evident in virtually every population surveyed.[32] The difference is not great, perhaps as little as one-tenth of 1 percent. However, it is real and it is consistent. With hundreds of thousands of babies being born each day, this differential can add up quickly. Moreover, sex ratios are not constant in any population. Under extreme conditions such as warfare, sex ratios can be even more skewed. But even under peaceful and what one may consider normal circumstances, the ratio of males to females fluctuates depending on the age class being considered.[33]

Subsequent to birth, mortality differences between boys and girls tend to be muted, but there is variation resulting from environmental and cultural factors. For example, among the Tsimane, a native population of forager/horticulturalists in Bolivia, the mortality rate among boys is higher than that among girls from birth to age fifteen. Although age-specific adult mortality has declined with acculturation and the availability of medicines and other resources, male mortality remains higher than female mortality. The primary cause of mortality during childhood among the Tsimane is infectious disease, suggesting that boys may be more susceptible to infection for innate reasons, because of greater risk, exposure, or both.[34] Female-biased infanticide is also a tragic source of sex ratio variation in many parts of the world. As children enter adolescence, disparities tend to become amplified with male mortality outpacing that of females. The source of this variation is both biological and behavioral. Indeed, one can make the compelling argument that biology influences the behavior of teenage boys and young men in a manner that increases their probability of an early demise.

THE BUMP

When I was twelve I believed I was Evel Knievel. For those not familiar with 1970s pop culture, Knievel was a daredevil who attempted to leap over anything that would result in a paycheck on his red, white, and blue Harley Davidson motorcycle, including school buses, the fountains at Caesar's Palace in Las Vegas, and the Snake River Canyon in Idaho. For the most part he was successful, but sometimes he was not. According to his biography, he broke virtually every bone in his body during his less than successful attempts.[35] My forays into daredevil stardom were not as dramatic but resulted in a number of spectacular failures. For example, I once managed to tear a silver-dollar-sized hole in my knee attempting to leap the span of my neighbor's lawn with the help of a rotting piece of plywood, a pilfered milk crate, and my brother's hand-me-down bicycle. To my knowledge none of the girls in my neighborhood suffered similar injuries. Yet after a visit to the local clinic and several painful stitches, my male friends and I shared many an afternoon comparing scars and neighborhood "war" stories.

Today the Internet is full of shows and videos that are almost entirely devoted to the pursuit of ever-increasing heights of male stupidity and males exhibiting a sharp decrease in risk aversion. Yet many of the exploits I watch as I eat my lunch at my desk provide me with a hearty chuckle. Why? Because in them, I see the inner core of virtually every male, including me. During their youth, males often discard good judgment and honestly believe that they are invincible, immortal, and downright bulletproof. Now well into my midlife years, I cannot get out of bed without worrying I might pull a hamstring.

The significant increase in male mortality during adolescence and early adulthood is well documented in a number of populations. Evolutionary psychologists Martin Daly and Margo Wilson were among the first to bring this issue to the forefront of human evolutionary biology. In their classic book, they showed how mortality spiked in young, urban Canadian males; their data were very similar to that for American males.[36] At first glance, one might assume that modern technologies might simply allow young males to be more effective at landing themselves in the morgue. Modern machinery such as automobiles and motorcycles, for example, may simply be amplifying our ability to kill ourselves.

There is a bit of truth in that assumption. Compelling evidence suggests that, since the industrial revolution, human male mortality has increased, most likely owing to the hazards of modern life. Anthropologist Daniel Kruger has done some remarkable work on this topic, drawing on various databases to illustrate changes and differences in mortality. Kruger has shown how excess male mortality compared to female mortality during adolescence and early adulthood virtually tripled in the United States between 1931 and 1995. While World War II obviously contributed to the increase between 1941 and 1945, along with the Korean and Vietnam wars a bit later, most of the increases are evident after 1981; mortality in young males was three times that of females.[37]

In human biology, this is indicative of a secular trend, that is, an increase in a human biological trait in response to changes associated with modern, industrialized lifestyles. This is most commonly observed in changes in height, weight, and age of menarche.[38] For example, my father and uncles were much taller than their father, who emigrated from Mexico and endured significant periods of hunger. Life for my father and uncles as copper miners in Arizona was no picnic, but food availability and life in general were a bit better compared to what my grandfather experienced. Fast-forward to my cheeseburger- and pizza-eating generation and it is easy to observe how my cousins and I tower over our parents and grandparents.

The research of Kruger and others is a grim reminder that modern conveniences also have the potential to create novel hazards that were simply not present during our evolution. So why don't we simply avoid going fast, wear helmets, and deploy these large brains that we conveniently evolved? The most salient answer is that there has been little or no selection to fathom the risks of riding on or in hunks of fabricated metal at speeds that were unthinkable only a hundred years ago. Traveling at a modest speed, say fifty miles an hour, was only possible if you leaped or fell off some high point. Needless to say, those who did seldom survived much less reproduced. One might argue with some degree of certainty that natural selection is still sorting this out.

So if young male mortality is affected by modern hazards, perhaps this is simply an epiphenomenon of the developments of modern, industrialized lives. Was the early male mortality bump evident in our hunter-gatherer ancestors? Without the benefit of a time machine, our best way of tackling this question is to examine contemporary hunter-gatherer populations. The demographic assessment of young male mortality requires a significant amount of raw data on deaths from a population, including (1) robust data sets, (2) large populations, and (3) more than a few intrepid researchers willing to collect the data. These are all in short supply for forager societies. Some of the most compelling data come again from the Ache since all three requirements can be met. Although population sizes are relatively small compared to modern societies, the longitudinal nature of the data allows for the fine-grained analysis that is necessary to detect significant changes in young male mortality.

Among the Ache, the male mortality bump is quite evident despite the lack of modern hazards. Information from the Ache or any other forager population is not perfect since we cannot assume that they are some sort of pristine population that is not affected by the outside world. Nonetheless, the adolescent mortality bump appears to be common among human populations, begging the question why this is so.

Evolutionary demographer Joshua Goldstein has argued that the declining age of the peak mortality during the adolescent bump is due to accelerated sexual maturation in response to improvements in environments, declines in infectious disease burdens, and energetic resources. This in turn allows males to become sexually mature earlier and increase sex steroid levels that facilitate risky behavior (see Figure 2.2).[39]

As alluded to by the model put forth by Goldstein, the underlying sources of higher male adolescent mortality are probably embedded in the biology, environment, and culture of being a male. In terms of biology, the neuroendocrine factors that facilitate risky behaviors never really go away after the "bump." The onset of high male mortality at young adulthood continues throughout a man's life and signals the onset of what is to be higher mortality and faster aging compared to women. Men still have much higher testosterone levels compared to women, even toward

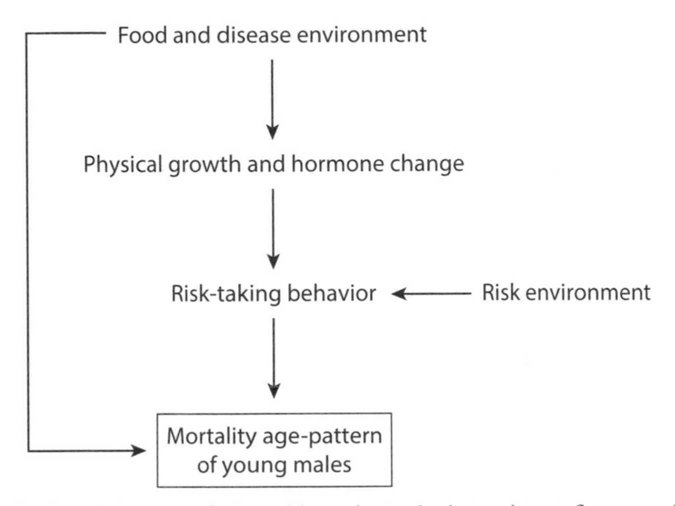

Figure 2.2. Causal diagram of selected hypothesized relationships influencing the male accident hump.

J. R. Goldstein, "A Secular Trend toward Earlier Male Sexual Maturity: Evidence from Shifting Ages of Male Young Adult Mortality," *PLoS One* 6, no. 8 (2011): e14826.

the end of life. Moreover, it may be useful to view the young adult spike in male mortality as a distillation of the factors that affect male mortality throughout their entire lives. An important reason why it might be intensified during the late teens and early twenties is that this is when women are at their peak fertility; during this time, males tend to adjust their priorities to take advantage of these potential reproductive opportunities. Safety and well-being are not so important.

A dramatic increase in male mortality usually coincides with an increase in reproductive effort, usually in the form of pursuing more opportunities to mate. In organisms such as deer that are seasonal breeders, high levels of mating go hand in hand (hoof in hoof?) with greater male mortality. This is also evident in many primate species. A small mammal that should serve as the poster animal for increased male mortality and mating effort is the Northern Quoll (*Dasyurus hallucatus*), a small, unassuming Australian marsupial that spends much of its life rummaging around the forest floor for insects and fruit. When the males of this species become sexually mature, everything changes. It becomes all about the mating process. During the course of their mating season, the males exert so much effort and invest so much energy in reproduction that they experience arguably one of the highest rates of male mortality of any mammal. They are the closest mammalian equivalent of salmon that die almost immediately after spawning.[40]

Many risky behaviors can also be classified as "display," that is, behavior that is meant to attract the attention of potential mates. Many studies show that younger women often prefer men who engage in some form of risky behavior.[41] Other data indicate that women prefer men who exhibit a hormonal milieu that is reflective of risky behavior.[42] Why is this? It is somewhat unclear, although some researchers have theorized that women are gauging the quality of men in the same manner that some birds do. In sage grouse, males display their breast pouches and plumage to females that observe and assess from the outside. It may also be that human females are assessing which males would be willing to put themselves at risk in the service of a female and her offspring.

Sometimes the daily tasks and behaviors that are expected of males happen to be ones that are more risky. For example, in hunter-gatherer societies, men often do most of the hunting (although there are some interesting exceptions), which is among the most hazardous of all tasks. Men step on snakes, get attacked by the very animals they are hunting, and in general are exposed to more dangers than are women who remain in the community. In this case, the sexual division of labor that is commonly seen in many societies contributes directly to greater male mortality. This is not to say that women's tasks are not without risk. However, the data do strongly indicate that men suffer greater mortality by engaging in the tasks that are expected of them.

But what happens when one enters the calmer elder statesman years? Eventually one trades in the motorcycle or sports car and purchases something that can accommodate an older lower back and slower reflexes. Male mortality remains higher than female mortality well into the later years of life. In the next section we explore mortality disparities at older ages when one might assume that the tables even out between men and women. As we will see, the differences in mortality have deep biological roots that transcend environmental influences and continue well into the later years.

WHERE ARE ALL THE DANCE PARTNERS?

As a population ages, men become scarce. A visit to virtually any retirement facility will reveal that there are many more women than men, and the mortality disparity between the sexes becomes even more apparent with time. After the age of sixty, the mortality gap between men and women widens, with men dying of age-associated illness at a much higher rate than women. Interestingly, and tragically, suicide rates in older men also jump in many societies. This demographic shift is common across all populations and, indeed, across most mammalian species.[43]

The reasons for greater male mortality at older ages are many. Some are likely to be rooted in sex-specific physiological factors such as hormones, metabolism, and possibly genetic disparities. If a man is lucky enough to outlive other males in his cohort, what benefits await him, and how can natural selection favor a man who suddenly finds himself surrounded by postmenopausal women? We need to understand that retirement facilities are recent constructions of modern life. It is doubtful that social arrangements that are common in retirement facilities were present during human evolution with cohorts of older individuals living together as social units. Older men who managed to outlive other men were probably immersed in a community filled with individuals of various ages of both sexes.

There is an interesting suite of differences between a seventy-year-old woman and a man of the same age. First, despite a likely decrease in somatic and possibly sperm quality, he is still able to father children. In other words, an older man does not undergo the same abrupt decline in fertility that a woman does. Is he as likely to father a child as he was when he was in his twenties? No. An assumption that older men are not fathering as many children as younger men is evident in demographic data that report metrics such as "reproductive value," an assessment of the probability of an age class to make a genetic contribution to the subsequent generation. The reproductive value of older men is lower compared to that of younger men, but it is not zero. An older man's probability of fathering a child is also lower. A comparative investigation of various forager populations revealed significant variation in male fertility, suggesting that assumptions based on Western populations are not universal.[44] The likelihood that an older man will father any children during his later years is primarily constrained by his ability to find a willing partner to bear his children. This happens more often than one would think, especially in societies whose customs and traditions are outside the Western norm of marriage and pair bonding. The significance of older male fertility on human longevity will be covered in greater detail in a later chapter.

Another difference between an older man and an older woman involves the question of paternity. At the end of my first book, I wrote a rather depressing phrase that was picked up by a book reviewer: "From an evolutionary perspective, men are quite alone." I still stand by those words. What I was trying to convey was that unlike women, men never know with absolute certainty whether their children and subsequent grandchildren are biologically theirs. It is a somewhat sobering thought, but it is true. Women know without a doubt who their children are since they bore them. Even in an age of modern reproductive medicine and artificial insemination, short of a paternity test, father-child genetic relationships are not as certain as mother-child ones.

Before the female readers and A-1 fathers out there send me hate mail, let me say that I readily recognize that there are many, many social and personal mechanisms that increase the probability of paternity, and for all intents and purposes, most men can sleep well at night knowing that the lights of their lives in the next room(s) are theirs and loved, no matter what anyone says. However, there are many situations in which paternity simply is not certain. The implications of this uncertainty have enormous ripple effects with regard to evolution by natural selection. One would predict that the sex that has greater uncertainty with regard to offspring relatedness would engage in less parental care. In humans and, indeed, most mammals, this is certainly true. However, human males engage in arguably the broadest range of variation in paternal care. Men can engage in behavior that is among the most devoted and caring to both mother and child but also descend into horrendous acts of violence and infanticide. Most other great apes do not engage in such a broad range of behaviors. In fact, with the exception of lesser apes such as gibbons and siamangs, no apes engage in any offspring care. Despite paternity uncertainty, men engage in a significant amount of care. Therefore, older men and older fathers have the capacity to contribute to child care as a way of increasing their fitness. This was a major development during the course of human evolution, one that had and continues to have a significant effect on our species.

But since evolution by natural selection relies on the propagation of one's genes, either through direct reproduction or investment in individuals who are closely related, such as grandchildren, nieces, and nephews, how can such paternal care evolve in men given the constraints that are evident with paternity uncertainty? Here we begin to enter the interesting and evolving arena of alloparenting. In essence, the idea is that for the energy- and care-intensive life history traits that are characteristic of humans to evolve, such as long childhoods, large brains, high fertility for an ape of our size, and long post-reproductive lives, humans had to have pooled their resources and efforts to support these expensive traits.[45] Some of those resources may have come from men providing far more food and calories to the community than what they themselves consume.[46] Older men among the Ache of Paraguay tend to peak in their hunting returns right around the time they start becoming grandfathers, in their mid-forties. This is somewhat conjectural but does resonate in explaining the advantages of outliving your male peers.

However, there are other potential benefits to others that may not be related. There is a vibrant area of research that suggests that humans may have cooperated in ways that were unique to the species, thereby supporting the evolution of our expensive life history traits. For this to occur, there would have to be extensive interaction between individuals who are unrelated. An interesting finding that was recently reported in the journal *Science* is the discovery that most individuals in foraging societies interact primarily with people who are not related to them. This means that contrary to common assumptions in biological anthropology, the influence of kin selection seems to be tempered by the fact that most people interact with those who do not share much if any of their genetic material.[47] Therefore, older men may have had the opportunity to contribute to the well-being of individuals who may not necessarily be family. This is an interesting idea that certainly merits greater attention. In other words, being nice to your neighbor is important, and paternity certainty may have become less important for men to evolve into caring fathers during the course of human evolution.

However, not every aspect of having an older male around is necessarily positive. Older individuals, regardless of sex, tend to require more care and resources. For those of us who worry about the care and tending of our aging parents, this is all too relevant. Who cares for older men who are likely to require more attention than women of similar ages? The archaeological record is filled with evidence of individuals with debilitating skeletal ailments and injuries that would have required significant care and tending for the individual to survive.[48] Clearly care is part of our evolution.

Interesting research from Poland has shown that paternal life span is significantly increased by the presence of more daughters, suggesting that they made significant contributions to the care of their elderly fathers. The number of sons had no effect on paternal life span, presumably since sons are away finding their own fortunes while daughters tended to stay close to home.[49]

This example illustrates a point that has yet to be brought forward: the importance of non-biological factors such as culture. Treating biology and culture as separate influences on male aging is unwise and likely to lead to incomplete interpretations of anthropological data. As my colleague Claudia Valeggia is apt to say, "Biology informs culture, culture informs biology." This is a basic tenet of biocultural research. The example from rural Poland shows how cultural variation can and does influence the expression of biological variables. Nonetheless, culture is constrained by our biology. We need to pay attention to both.

Leaving culture aside for the moment, we need to reengage with biology and discuss the physiological changes that occur in older men. Demographic changes in mortality and morbidity are rooted in how our bodies change over time. Specifically, older men undergo bodily changes that define male aging. Whether one lives in the jungles of South America or an eastern European city, changes in body composition are common aspects of male aging. Time to get a handle on love handles.

CHAPTER 3

GETTING A HANDLE ON LOVE HANDLES

> They were big powerful men, with not much capacity to weigh the consequences, with courage, with strength, even yet, though their skins were no longer glossy and their muscles no longer hard.
>
> —Joseph Conrad, *Heart of Darkness*

I hate sit-ups. When I was younger I sported a more or less flat abdomen with just some modest effort. As I write these words on my laptop, I'm hesitant to glance down from the keyboard and see the middle-aged man paunch that has creeped into my life. Thankfully it's not that bad, but it's there. I admit that I don't exercise as much as I did in my early years, but it is clear to me that my body is not as forgiving as it was to resisting the emergence of love handles and a not-so-flat abdomen. Fat has replaced muscle.

Here I will argue that men have adjusted to the bodily changes caused by aging in unique ways compared to other great apes and mammals in general. The ideas are not ironclad, but I do think they merit serious consideration. In many ways, the adaptive changes made by men are congruent with the overall ability of humans to be extremely malleable in response to environmental challenges—or, in this case, aging-induced challenges. Organisms can adjust their behavior and to some extent modify their biology to cope with the effects of age-associated degradation. Although these efforts ultimately only delay death, they do provide organisms with additional opportunities to increase their fitness at older ages. One might consider this to be turning evolutionary lemons into fitness lemonade.

In this chapter, we will discuss how the basic ingredients of a man change with age, how body composition is modified with age, and how men may have leveraged these changes to increase their fitness in later years. Many older male readers may find themselves nodding in somewhat grim agreement, I hope with at least a slight, chagrined smile. We will also take a good look at metabolism, or the manner in which men process energy, and the hormonal aspects of male aging since these are key components of energy management.

SOFT IN THE MIDDLE

Erasmus Darwin was not as famous as his grandson Charles. Besides being the paternal grandfather of arguably the most famous naturalist of all time, he left his mark on history in several ways. As a wealthy English gentleman of considerable social standing, he was a progressive for his time and was well-known for his abolitionist views and for an early proposal that evolutionary processes were a driving factor behind the emergence and development of life on earth. It is thought by some that these ideas likely seeded the thinking of young Charles and ultimately led him to formulate his theory of natural selection, the evolutionary mechanism that had eluded his paternal grandfather. Erasmus also left other marks, particularly at the dining table: he was so corpulent that by the time he was in his late forties, a semicircular cut had to be made in the table to accommodate his significant girth. Likewise, the father of Charles, Robert Darwin, was so portly that Charles commented that when he entered the room it was like the tide coming in.[1]

The girth of many older men tends to grow a bit over the years, although thankfully most dining tables remain intact. We have all bought and donated our share of casual slacks, although one pair holds a special place in my closet. I won't divulge the brand, but suffice it to say that I call them my stretch-o-matics. Deftly hidden in the waistband of a dark gray pair of casual slacks is an elastic waistband that allows me to fit into these pants comfortably as I fluctuate between a waistline of 34 and 36, usually in response to the long Connecticut winter and my need to get into some degree of physical shape when I'm about to head to the Amazon to conduct field research. While I readily admit that I'm a bit embarrassed to write these words, I know I'm not the only one. You know who you are. However, in the interest of science education, I'm willing to allow my ego to take the bullet. Some of you might justifiably ask why I don't get off my rear and exercise more.

I also canceled my gym membership. Mostly because it was money wasted since I'm not much of a morning person and I never had the chance to get there before they closed. When the New England weather cooperates I make do at the local high school running up stadium steps with my field backpack laden with bags of (unused) cat litter. It is much cheaper and effective for the type of things we do in the field. I also enjoy watching the woodchucks on the nearby hillside foraging and doing woodchucky things. But I have to confess that any disappointment in canceling my health club membership was tempered by my growing frustration that it took twice the effort to get a fraction of the fitness result that I used to enjoy when I was in in my twenties. When I was younger, I would lift weights, run, play pickup basketball, and afterward feel fantastic. Now in my fifth decade of life, I do feel better when I exercise and appreciate that it is still good for my cardiovascular well-being as well as my other bits and bobs. I never regret a good workout, but there is always a bit more soreness, a bit more time to recover, and more effort required to fit into my jeans. My body has changed.

For men these changes are very important since there is ample evidence that physical strength and prowess were very important during our evolution. However, as we will note in just a bit, strength is but one factor that is important to human males. The ability to run, throw, and do all the things that are necessary to be a successful hunter are all tied to some degree of physical strength, muscle tone, and overall somatic integrity. For men, many strength-dependent tasks are linked to physical traits that were unique to males, beneficial for male-male competition and/or attracting females. These are known as sexually dimorphic traits. In other animals, claws, fangs, and antlers are examples of such traits that are commonly deployed. Humans have relatively wimpy teeth and are devoid of claws or anything remotely sharp for that matter. Instead, males have had to rely on physical traits like upper-body strength and activities that are somewhat male biased, such as throwing projectiles. An excellent body of research by Neil Roach of Harvard and his colleagues has demonstrated this by examining our ability to use our shoulders to throw projectiles. Chimpanzees and other primates are also pretty good at throwing things, but there seems to have been selection for humans to hurl objects more effectively and efficiently. The musculoskeletal morphology of throwing seems to have evolved around the same time that hunting did. Coincidence? Perhaps, but probably not.[2]

Changes in body composition and a decline in physical efficiency are part of the process of aging. But why, for example, is it so difficult to put on muscle but so easy to put on fat? For some mammals and other vertebrates, the deposition of fat is crucial for survival. Penguins and cetaceans put on fat to make it through periods of fasting as well as to aid in buoyancy. Women are more efficient at putting on fat tissue compared to men in order to support the significant metabolic costs of pregnancy and lactation. In men, is the shift from muscle to fat adaptive or a constraint dictated by aging?

It may be a bit of both. An ample supply and the ready availability of cheeseburgers and such do contribute to the predicaments of many men. The fact that I sit at a desk for hours at a time also does not help. However, I know that even when I cut out the chili dogs and the more than occasional pint and force myself to take a walk around the block instead of reaching for another cup of coffee, my body simply does not respond like it did in the past. Women have very similar experiences, and it would be inaccurate for me to claim that this is a "male thing." However, the loss of muscle is more evident in men and arguably more detrimental in light of the tasks that were probably crucial for male fitness during human evolution.

MUSCLE TO FAT

Muscle is the predominant tissue in men. However, muscle is important to all vertebrates and invertebrates, both male and female, since it allows us to move about the environment. In males muscle serves two purposes that are unique to our sex. Aside from movement, it augments our ability to reproduce by supporting competition and attractiveness. It is also an important source of overall energy regulation. More so than other types of muscle, skeletal muscle is sexually dimorphic, which means that relative mass, form, and function differ between men and women. Sexual dimorphism in body composition emerges largely as a result of hormonal influences that take hold when both males and females begin to reproductively mature. Adolescent males build additional muscle tissue while young women deposit fat tissue in association with the onset of fertility. Women also have a significant amount of muscle, but men tend to have more of it, have different types of it, and have it in different places. The type of muscle men tend to have is type II, which supports quick movements and bursts of strength. The distribution of muscle in men is also different than it is in women. Men tend to have more muscle around the upper parts of their body, in their shoulders, arms, and back.

Besides being used for movement and strength, skeletal muscle is a repository of energy that is mobilized during periods of energetic stress. If you try to lose weight, it is a delicate balance of ditching the fat while maintaining muscle since your body wants to conserve energy by getting rid of the expensive stuff. For example, during my postdoctoral work at Massachusetts General Hospital, we conducted a pilot study on Harvard University rowers, who are divided between lightweights and heavyweights. The heavyweights were not constrained by weight limitations. They ate and trained like possessed fiends in their quest to get stronger. The lightweights, on the other hand, were constrained by weight limits on their boat and had to be very weight conscious and careful about lifting weights and watching their diet. They trained extensively but were mindful of their caloric intake. It was very difficult to control their weight without losing muscle.

In men, skeletal muscle accounts for about 20 percent of basal metabolic rate (BMR), which is a general measure of how much energy the body uses during metabolism, making skeletal muscle relatively expensive compared to other tissues such as fat, with the exception of the brain.[3] Although individual muscle cells are not very metabolically expensive, the sheer mass of muscle tissue in men makes it a rather pricey metabolic investment. To cope with this demand, skeletal muscle can catabolize or break down to release amino acids that can be deployed toward energy needs, sort of like burning your furniture to heat your house. Not optimal, but it works. But the catabolization of muscle tissue also decreases somatic metabolic costs during lean times or periods of intense immunological or psychological stress. The hormone cortisol, which is released to increase in cellular glucose uptake during periods of stress, is a primary culprit, promoting muscle tissue breakdown by blocking the anabolic or muscle-building effects of testosterone while promoting catabolism. In this sense, compared to females, skeletal muscle is a particularly vital energy regulatory tissue in men.[4]

One can hypothesize that the selection for this increase in expensive muscle mass was due to form and function. Females likely preferred men with this muscle distribution, and men with that distribution had a selective advantage over those who did not. The evolutionary biology of sexually dimorphic muscle tissue is not terribly complicated. All that is basically needed are testosterone and growth hormone receptors on muscle tissue to allow them to respond and grow when men produce anabolic steroids and peptide growth hormones. In other mammals, sexually dimorphic muscle tissue lies in areas that are more species specific and appropriate. For example, in deer, bucks tend to have sexually dimorphic muscle tissue in their necks in order to deal with antler duels with rival males.

Has natural selection dealt men a bad hand, allowing their bodies to waste away when they are finally starting to figure out how life works? Since it happens to other mammals, it certainly does not seem to be specific to older human males. What seems to be different is that while men's bodies start to betray them, behavioral and other somatic advantages appear to have been leveraged in ways that are unique to humans. It seems that some of the advantages that come with advancing age start to pay dividends after men have passed their physical prime. It turns out that as men age, lose muscle, and gain fat, all is not lost. But first it would be helpful to know what hormones make men get soft in the middle.

IT'S YOUR HORMONES

Declines in testosterone, lower metabolic rates, and shifts in other areas of the hypothalamic-pituitary-testicular (HPT) hormone axis are largely to blame. In order to keep muscle in place, one needs hormonal support as well as energy in the form of glucose and other sugars to keep muscle cells vigorous and functioning.[5] As men get older, lower testosterone levels contribute to this process, although there is a lot of variation between men in this regard. Among Ache men as well as men in central Africa, and Nepal, the level of testosterone does not differ very much with age, suggesting that the drop we see in places like the United States is unique to our particular ecology and lifestyle. Men living urbanized, sedentary lifestyles exhibit a more notable drop in testosterone compared to other men around the world,[6] resulting in equally significant changes in body composition.

However, many of the populations in these studies tend to be remote and may reflect conditions that are just not that common anymore. In a collaborative study that I conducted in Japan among healthy men, we found that testosterone tends to peak in the second decade of life and then slowly drop. Interestingly, though, testosterone ceases its decline after the age of forty and is pretty much stable for the rest of a man's life. My collaborators and I hypothesized that testosterone levels can be maintained in men who have strong social and family ties and exhibit relative health in other areas.[7] If you recall from the study in Poland, fathers with daughters had longer lives likely because of the care that they provided. Similarly in Japan, men with supportive and caring family members likely had better access to health care, which has been shown to promote higher testosterone levels with age.[8] There is also the psychological side of testosterone. The Japanese men in our study were a somewhat self-selected group who were able to travel to the community center to participate in the activities available there, including this investigation. They were surrounded by doting, supportive family members and generally treated with much care. Those who were too ill or who lacked family support were largely absent from the study. Social isolation with age often leads to depression, which is associated with lower testosterone levels.[9] The Japanese men in our study represented the healthiest and most socially engaged. Considering the importance of care, respect, and family engagement with the elderly in Japan, this would likely be an important factor in male aging within this society. Again, the value of a biocultural perspective gives us a knowing nod. Cultural variation and social interaction matter.

While testosterone levels exhibit some degree of decline depending on the population, there seems to be less of a change in muscle tissue's ability to respond to hormonal stimulation. In other words, muscle cells seem to maintain their ability to respond to testosterone if the opportunity arises. Endocrinologist Shalender Bhasin and colleagues compared the response of healthy younger (19–35 years) and older (60–75) men to testosterone supplementation. In order to get a hormonal "clean slate" they first blocked endogenous hormone production at the level of the hypothalamus and inhibited all endogenous testosterone secretion. They then replaced that testosterone in a graded dose dependent manner, all the while controlling dietary content, caloric intake, and activity. Surprisingly, increases in overall fat-free mass, muscle growth, and strength to graded doses were not significantly different between younger and older men. Both younger and older men responded positively in the same manner.[10]

An interesting side note to this study is how younger and older men responded to the same testosterone doses. Older men exhibited significantly higher testosterone levels in response to each graded dose compared to younger men. In addition, the level of luteinizing hormone (LH), the pituitary hormone that is responsible for endogenous testosterone production, was also consistently higher in older men. What does this mean? Younger men may have more efficient liver and kidney function, resulting in higher clearance rates, or, more interestingly, there may be differences in hypothalamic function in response to testosterone. In other words, despite declines in testosterone, older men maintain the ability for anabolic responses to testosterone. Why, then, does testosterone decline with age in some men? The answer perhaps lies in a disconnect between the male hormone production system and muscle receptors, revealing differential aging between these aspects of male physiology.

There is limited information about testosterone, reproduction, and so forth in similar studies on forager or horticultural populations. Anthropologist Benjamin Trumble has been conducting research on male reproductive ecology among Tsimane horticulturalists in the tropical regions of Bolivia. Trumble assessed changes in salivary testosterone levels among Tsimane men in association with anabolic activities and found that testosterone levels increased on average 48 percent in response to an hour of tree cutting. The effect was independent of age. Moreover, similar to other non-Westernized populations, the Tsimane exhibit lower baseline testosterone levels compared to Westernized men.[11] This suggests that older men retain the potential to increase testosterone levels when anabolic opportunities arise or when the metabolic demands of activity are required.

When inactive, older men seem to ratchet down their testosterone levels, perhaps to avoid the metabolic cost or as an unavoidable response to aging. The bottom line is that changes in body composition in response to changes in hormone levels do not seem to be an inevitable response to aging.

Another interesting change in male hormone function is diurnal secretion. In younger men, testosterone levels are highest in the morning and then wane in the evening. The reason for this diurnal pattern of secretion is a bit unclear but probably is embedded in the need to support activity after a night's sleep. However, this diurnal pattern diminishes with age. This is true for both American and forager men, suggesting that this is a universal phenomenon regardless of environmental influences.[12] Since other hormones, like cortisol, exhibit similar changes,[13] the underlying cause of this decrease in diurnal function probably lies in a decline in the function and sensitivity of the hypothalamus to energetic or circadian cues. In other words, a deep-seated change with age seems to be a declining neuroendocrinological ability to detect changes in the environment. In a species that excels at adapting to environmental change, this would seem to be an important detrimental effect of aging.

WHY THE DECLINE?

If men were to replace the hormones that they lose with age, they could regain the capability of building muscle and losing fat. Indeed, if you have ever browsed through an in-flight magazine, I am all but certain that you have encountered an advertisement by a certain physician who is a great proponent of anabolic hormone therapy for men. His regimen of hormone supplementation and exercise does appear to garner results, although I am not aware of any clinical studies indicating that this therapy regimen is safe. Nonetheless, other clinical studies strongly suggest that men can improve strength and vigor with hormones. But this raises a Medawar-like question. If taking hormones at older ages can promote vigor and presumably increase or at least maintain reproductive fitness, why hasn't natural selection allowed all men to look like a younger, more physically fit man using one's own hormones?

The most likely answer is one that accounts for energetics and interactions with other organ systems. It is unlikely that human populations during our evolution had access to the amount of calories and selection of food sources that we have now. This is a central reason why there are often discrepancies between results obtained from different populations.[14] In other words, even if a man has the capability of producing higher levels of testosterone and growing more muscle, he still has to feed that tissue. Even if you can buy a Ferrari, you still have to maintain it, the costs of which can often supersede the initial purchase price. Also, muscle-induced increases in metabolic rate can tax the limits of other organs such as the heart, liver, and kidneys. These tissues are also aging and becoming less efficient. Burdening them with the additional metabolic costs of muscle tissue may simply not be sustainable.

Like testosterone receptors in muscle, gonadotropin releasing hormone (GnRH) producing cells within a part of the brain known as the hypothalamus also seem to have a finite life span and lose sensitivity. What is GnRH? Briefly, GnRH is secreted in the hypothalamus within the brain and acts as the ignition for the rest of the reproductive hormone system in both men and women. Without it, reproductive function is a nonstarter. In other words, as men get older the signal is functional but the antenna within the hypothalamus isn't listening. Therefore, if an aging hypothalamus loses sensitivity to important hormonal cues and the ability to produce GnRH, endogenous testosterone levels will decline even if testosterone receptors in muscle and other areas retain their integrity and ability to function. In many ways it's as if the home office goes on vacation and leaves the rest of the workforce with nothing to do.

An important example of this decrease in sensitivity is a decline in the ability of the hypothalamus to assess environmental cues, specifically receptors for insulin that provide information on an individual's energetic status.[15] Insulin is a very good indicator of circulating glucose levels. We know that if you give a healthy man a bit of maple syrup, his LH and testosterone levels will perk up.[16] If a man fasts for a few hours, the response is even more noticeable.[17] But with age, hypothalamic sensitivity and perkiness to energetic status seem to diminish.

To understand this more thoroughly, it would be great if we could measure GnRH levels in association with age and energetic status. The challenge is that it is all but impossible to directly measure GnRH since it is only found in the tiny hypophyseal port that connects the hypothalamus to the pituitary.[18] One would have to drill a minute and exact hole into a subject's forehead to get a sample. We need to find a less intrusive research strategy. Instead, we can observe the effects of exogenous GnRH on downstream hormone levels and function in association with various experimental conditions.

Researchers led by the clinical endocrinologist Johannes Veldhuis observed changes in LH levels in response to GnRH administration in younger and older men undergoing mild fasts. Bear in mind that LH is the primary hormone that stimulates testosterone production in the testes. Make subjects a little bit hungry and one would expect the hypothalamus to respond to a bit of food. Instead of merely looking at hormone levels, they investigated the pulsatile patterns of LH secretion since hormones under the control of the hypothalamus are secreted in pulses, similar to a heartbeat. Therefore it is likely that similar to radio signals, pulsatility and amplitude will have an effect on target receptors and tissues. They found that older men exhibited no significant differences in LH pulsatility between fed and fasted states whereas the younger men exhibited a slower and less vigorous pattern of LH pulsatility. When younger men were given GnRH in the fasted state, their LH levels jumped and they resumed more vigorous and frequent pulse releases of LH. The home office was open for business. In contrast, the older men did not respond to GnRH and continued to secrete LH in a more constant and less vigorous pulsatile pattern.[19] The home office was not answering the hormonal phone.

So what does this mean? One could interpret these results as evidence that older men become less sensitive to environmental and energetic cues, become less malleable and responsive to surrounding changes. This loss of hormonal plasticity causes the body to become less efficient at putting on muscle and regulating fat accumulation. In essence, the brains of older men stop receiving clear information on the environment and energetic status. This may be a core aspect of aging that could contribute to other conditions such as weight gain and muscle wasting.

LOWERING THE FURNACE

You are such an endotherm. Yeah, I said it. You're an organism that generates its own body heat. After you account for the energy needed to process your lunch, nutrients from your food are either burned for immediate needs or stored for future deployment, usually in the form of fat in adipose tissue. Overall, if you recall from a bit earlier in this book, the production of energy in tissue is metabolism. Anabolism, or the building of new metabolically active tissue, is usually facilitated by hormones such as testosterone, growth hormone (GH), and insulin-like growth factors (IGF). Again, catabolism is the breaking down of tissue into its component molecules, either to liberate resources for metabolism, as in the case of catabolizing fat, or to decrease the amount of tissue that needs to be maintained. Although these processes are very similar between men and women, the extent and effects on reproduction are often unique to each sex.

For example, cortisol is a hormone that facilitates the cellular uptake of glucose. Levels of cortisol usually increase in response to attacking Chihuahuas, math tests, and anything else that promotes the impulse to run away in a cowardly fashion. Listen to the lyrics of "Gimme Three Steps" by Lynyrd Skynyrd and you'll know what I mean. Cortisol also affects reproduction in men and women in different ways. In stressed-out men, high levels of cortisol can block the production of testosterone and increase the probability of sexual dysfunction and increased muscle catabolism. In women, high levels of cortisol can perturb pregnancy; this can affect fetal growth, produce shorter gestation and early onset of labor, and increase the likelihood of fetal loss.[20]

Also during pregnancy, the average energetic demands of fetal boys appear to be greater than they are for fetal girls. Although it is difficult to measure fetal energetic expenditures, we do know that the caloric intake of women carrying male fetuses tends to be almost 10 percent greater than it is for women carrying female fetuses. This seems to account for the higher birth weights in males.[21] These differences in total metabolic costs continue through childhood and into adulthood.[22] However, it is important to note that metabolic needs are not static and increase dramatically in women during pregnancy and lactation. For men, metabolic rates are closely tied to sexually dimorphic lean body mass. Therefore, comparing average energetic demands between men and women is compromised by many nuances that need to be taken into consideration.

As both men and women age, metabolic differences become less pronounced between the sexes, mostly due to changes in body composition in men. But men's cells also change with age. Although energy expenditure differs greatly among various tissues, BMR is a sort of average of how much energy it takes to keep body and soul together. Very roughly, human males need between 2,000 and 2,200 calories on a daily basis just to stay alive and keep their weight constant.[23] Besides the brain and various internal organs, a major contributor to BMR is skeletal muscle mass, which is important since, as we've noted, skeletal muscle is the most sexually dimorphic and unlike fat tissue contributes the most to overall BMR in men. Although each cell in itself is not very costly compared to, say, a brain cell, the sheer number of skeletal muscle cells results in a very significant contribution to male BMR. In general, as men lose muscle mass, they exhibit lower BMR and energetic expenditure.[24] Since most men living under conditions of energetic surplus (i.e., urban American men) do not pare down their food intake, there is more excess energy and greater fat deposition. Hello, stretch-o-matics.

While body mass and composition differences among Westernized populations have been well documented, only recently have we obtained metabolic data from contemporary foragers. A recent analysis of resting metabolic rates among Hadza foragers and those living in the city using doubly labeled water, the gold standard of resting metabolic rate (RMR) measurement (similar to BMR but without the absolute quiescence), showed that Hadza men expended 2,649 ± 395 (average ± standard deviation) kilocalories per day compared to 1,877 ± 364 for women, a whopping 41 percent difference. Though the sample sizes were modest (13 Hadza men and 17 women), the results are compelling.[25] As a researcher who spends much of his summers lugging portable freezers, solar panels, and motorcycle batteries around the Amazon to conduct related research, I would be remiss if I did not remind the reader that the logistical challenges of conducting such research in the remote corners of Tanzania are considerable. While physical activity levels were higher among Hadza hunter-gatherers compared to those of farmers and Western controls, daily energy expenditure was about the same, suggesting that energetic expenditure is a conserved trait that is relatively invariant in response to cultural variation. Preliminary results from Shuar forager-horticulturalists in the Ecuadorian Amazon reveal similar findings.[26]

These studies also showed that sex differences in energetic expenditure are robust and significant, resulting in a significant amount of male excess calories to be expended over a lifetime. Does this contribute to higher adult age-specific mortality in men compared to women? I would postulate "yes." One can test this hypothesis by experimentally adjusting energetic expenditure in males and females. Why would males invest in this cost if it does not improve fitness? Since sexual dimorphism is largely a function of males investing in tissue and other forms of biomass that are beneficial for competitive ability, attraction, and other aspects, sexually dimorphic biomass in males can be viewed as metabolic investment that reflects reproductive effort.[27] Results from nonhuman animal experiments support the hypothesis that male-specific metabolic rates do contribute to higher mortality, most likely due to the burning of fat resources and a compromised ability to cope with energetic stress.[28]

Not surprisingly, the age-associated decline in energetic expenditure in men is primarily due to changes in body composition. As muscle mass declines with age, fat deposition increases, which also contributes to a decline in BMR since adipose tissue is metabolically inexpensive.[29] Decreases in metabolic rates seem to be pretty common in organisms since even fruit flies exhibit a drop with age, usually within the first two weeks of life.[30] Increases in male adiposity with age compared to that of females are not uniformly distributed around the body. When looking at subcutaneous compared to visceral fat, men tend to grow around the midsection (visceral). The accumulation of this fat is the result of lower testosterone levels and increases in estrogens.[31]

However, the picture gets a bit more complicated when population variation in adiposity and hormonal regulation of fat metabolism is brought into play. For example, leptin is a hormone that is secreted primarily by fat cells and serves as a signal of fat availability to the hypothalamus. One can imagine leptin as a kind of bank statement that keeps the hypothalamus informed of available energetic balances. Leptin affects a multitude of functions, including food intake, metabolism, immune function, and some aspects of reproduction.[32] Non-Western populations exhibit leptin levels that are low compared to those of Westernized populations.[33] Moreover, it appears that leptin is not a robust signal of adiposity in non-Western populations and, interestingly enough, chimpanzees. In both cases, leptin is poorly correlated with adiposity in males, although it is clearly associated with fat in Westernized males.[34] The role of adiposity in male reproductive effort remains to be clarified, although clearly hormonal mechanisms that are central to the function of adiposity are subject to plasticity and variability.

What this means in non-endocrinological plain speak is that the regulation of fat tissue may have been a recent evolutionary phenomenon in human males. Growing and shrinking muscle through the regulation of testosterone and other anabolic hormones was the primary way in which men adjusted to environmental stresses and food shortages during our evolutionary past. The fact that well-fed male chimps are pretty much deaf to leptin except in extreme cases of obesity indicates that hominid males managed their energetic budgets via muscle and anabolic hormones. Pizzas, bread, and other goodies that promote fat deposition seem to have nudged their way into the picture only recently. For us WEIRD older males, our environment, diet, and lifestyles have complicated our hormonal lives. Growing up WEIRD not only promotes fat deposition but complicates our hormonal environment in a manner that is unique to our evolutionary history.

I'VE BEEN AROMATIZED

Testosterone tends to receive much of the hormone press when it comes to male aging. Others such as human growth hormone (hGH) also vie for the spotlight, but testosterone tends to get the lion's share of attention. However, there is another hormone that has not received its due and is directly linked to male aging, fat, and body composition: estradiol. Readers with some degree of familiarity with hormones may be thinking that this is a typo. Estradiol, a primary estrogen, is a female hormone. What does it have to do with men? Well, last time I checked, hormones are not produced with male or female labels. The reason testosterone has garnered so much attention is because men produce so much of it compared to women and because of its proven role in growing and maintaining muscle mass, tweaking libido, and changing boys into men. However, it only takes one enzymatic change to convert testosterone to estradiol. The enzyme that can do this is aromatase. Remember that name. I all but guarantee you will be hearing more about this enzyme in the years to come, probably when testosterone patches fall out of favor. But I digress.

Aromatase is an enzyme that is coded by the gene CYP19 and is metabolized in a number of tissues including those in the brain, testes, and, most important for our discussion, fat cells. The function of aromatase is to transform testosterone into estradiol. A common expression of the role of aromatase can been seen in obese women. How so? Aromatase is commonly found in a part of the ovaries called granulosa cells. Adjacent to granulosa cells are theca cells that create testosterone in response to LH from the pituitary. Interestingly, these cells also respond to insulin by increasing testosterone production. In obese women with insulin resistance, insulin reaches levels that stimulate an overproduction of testosterone in the ovary, much more than what can be converted (aromatized) to estradiol. Consequently, excess testosterone spills over into general circulation and contributes to many of the secondary symptoms seen in obese women with metabolic syndrome, including head hair loss (alopecia) and unusual hair growth on the face, arms, and body (hirsutism), places where men commonly have hair.

Men who exhibit a variant of the CYP19 gene that stimulates increased aromatase activity exhibit a greater incidence of gynecomastia (breast growth).[35] However, the most common effects of aromatase are evident in obese men. Men with an overabundance of fat cells (adipocytes) tend to exhibit symptoms that can ultimately lead to what clinicians dub hypogonadism, that is, a lack of testosterone that leads to gynecomastia, shrinkage of the genitals (microphallus), and a continuing decline in testosterone. Weight loss and the elimination of adiposity can increase testosterone levels and decrease estradiol through the elimination of adipose-driven aromatase activity.[36] As men get older and their metabolisms slow, it becomes much easier to lose muscle mass and gain fat. This is a positive feedback loop since greater deposition of fat leads to more aromatization of testosterone, higher estradiol levels that facilitate fat deposition, and a negative feedback effect on GnRH levels by increasing levels of estradiol that can further suppress the reproductive hormone system in men. In other words, aromatase increases the female hormone factor while decreasing our maleness.

It is interesting that in lean hunter-gatherer men, estradiol levels do not increase with age but are positively associated with testosterone, suggesting that adipose-driven aromatization does not play a major role. Most estradiol tends to be produced in modest levels locally in the testes, which is often the case in lean men.[37] However, this is not so in Western, sedentary men with more than a few fat cells. Indeed, it can be argued that the obesity epidemic might underlie some of the reports of low testosterone levels in men and perhaps the thriving market in male hormone replacement therapy.

WHY NOT TURN UP THE HEAT?

If decreases in metabolic rate with age are detrimental to the ability to put on muscle, to attract mates, and to maintain a youthful vigor that would promote reproductive success, then why hasn't natural selection cobbled a solution together? The likely reason is that there is a cost—or, more accurately, a cost that cannot be endured anymore because of aging. Recognizing this cost requires us to understand the significance of the differences in energy invested in reproductive effort between males and females. In mammalian females, the metabolic costs of reproductive effort are pretty straightforward. Pregnancy and lactation both involve metabolic costs that are directly relevant to reproductive effort.[38] In males, the metabolic costs associated with reproductive effort are tied primarily to sexually dimorphic tissue.

Anthropologists Caroline Key and Catherine Ross examined the metabolic costs of sexual dimorphism in a number of primates and found that when male body size exceeded that of females by 60 percent, the metabolic costs in males were greater than the reproductive costs in females.[39] Only the most sexually dimorphic primates would be affected by this cost. However, this analysis does not take other sources of energetic investment into consideration, such as provisioning and caretaking activity, costs that are more commonly borne by human males. In addition, if investment in sexually dimorphic tissue can increase fitness, there must be a constraint or trade-off that is tempering the benefits of greater muscle mass and higher metabolism. In males, the most common life history trade-off is between reproductive effort and mortality.[40]

Is there a relationship between BMR and mortality across ages in men? There is evidence to suggest that there is. To date, there is only one study that has examined this in a comprehensive manner. Carmelinda Ruggiero and colleagues looked at common variation in BMR in 972 men over a forty-year span as part of the longitudinal Baltimore Men's Study, arguably the longest and most comprehensive study of men's health to date. Their findings are derived solely from an American population but are nonetheless quite interesting. First, BMR declined with age, which was no surprise. However, when they looked at variation in BMR at older ages, it seemed that men with naturally higher BMR had a greater risk of mortality. They found that compared to the reference age group of 31.3–33.9, elevated BMR in the second age group (33.9–36.4) was 28 percent higher, and in men over the age of 36.4, the risk of mortality was on average 53 percent higher if the metabolic rate was elevated.[41]

Women also participated in this study, but there were no similar associations between higher BMR and mortality as were seen in the men. Other co-factors that were controlled were smoking, diabetes, cancer, and various other physiological measures. The authors conclude that high BMR may be a marker of poor health and therefore greater risk of mortality, which may very well be the case. It has been established that infections significantly increase BMR, so infectious status may be at play.[42] Nonetheless, this study shows that elevated BMR is deleterious to survivorship in older men. However, this study raises the question of why some men have higher BMR than others. Infectious disease may play a role, but there may be other factors that remain to be identified.

It has long been suspected that males' higher energy demands are linked to mortality and longevity differences between males and females.[43] As we will see, the role of energetic and metabolic differences in male mortality is influenced by a number of factors. However, the predictions and well-established theoretical foundations of life history theory strongly suggest that energetic and metabolic differences between males and females should contribute to differences in mortality patterns. Higher male mortality due to differences in metabolism is likely linked to two aspects. First, since males have higher energy demands, they are more susceptible to trade-offs that would require them to sacrifice energy allocation toward maintenance and thus survivorship in the face of other needs, particularly reproductive effort. Second, the greater energetic demands of males may indicate that downstream costs caused by toxic byproducts of burning calories, especially in the presence of oxygen, lead to accelerated aging in men compared to women.

EVERY BREATH YOU TAKE

As a younger person in the early 1980s, I was among the hordes that rocketed the song "Every Breath You Take" to the top of the charts, buying the vinyl 45 single as soon as I could scrape the money together. The Police (as Sting and his lads were formerly known) not only wrote a very catchy love tune, despite the stalker overtones, but also inadvertently tapped into a very important aspect of aging biology—oxidative metabolism—since literally every breath you take contributes to the aging process. Oxygen is necessary for our very survival and allows for more efficient energy usage through aerobic metabolism compared to anaerobic metabolism, which does not require oxygen. But as with every other aspect of being a living organism, there is a cost and trade-off. The relationship between oxidative metabolism and the aging process is complex, but several salient factors link this aspect of our biology to aging.

Aerobic metabolism generates toxic by-products. Think of what comes out of the tailpipe of your car. When your car burns fuel, it creates waste, some of which is toxic. The same thing happens in your cells. Besides carbon dioxide, your cells create toxic agents that contribute to oxidative stress. These agents are in the form of reactive oxygen species, peroxides, and free radicals. When you take vitamins, gulp down a handful of blueberries, or dine on a nice piece of wild-caught salmon, you are taking in antioxidants that may help ameliorate the effects of oxidative stress agents. Oxidative stress is among the most basic processes that are believed to contribute to aging in all organisms that breathe air, which is pretty much almost everything. As we discussed earlier, body size and metabolic rates contribute to rates of senescence, and there is compelling evidence to suggest that body size and metabolic rates are associated with oxidative stress. That is, larger organisms tend to exhibit lower rates of oxidative stress compared to smaller ones because of their slower metabolic rates, but there are numerous factors that can affect this relationship.[44]

Let's start with reactive oxygen species (ROSs). When a cell in your body uses a molecule of oxygen to help create adenosine triphosphate (ATP), the basic unit of energy, it creates toxic by-products including superoxides. These are molecules comprised of two oxygen atoms and a spare electron, resulting in a strong negative charge. This negative charge is called ionization. This causes the superoxide to be as clingy as a nylon sock fresh out of the dryer and just as attractive to most other molecules. Sometimes this is a good thing since the production of superoxides by the some immune cells is the weapon of choice for dispatching foreign pathogens during infection. However, when produced as by-products, superoxides tend to wreak general biochemical havoc if unchecked. In other words, they are quite toxic.

Millions of these superoxides are created every second as your cells metabolize energy. Fortunately, protective agents created by your body mop up most superoxides. Biochemical protective agents include those with exotic-sounding names like superoxide dismutase (SOD). Say it loud enough with your fist raised high and you will look and sound like a comic book superhero. SODs sop up superoxides and either render them harmless or alter their chemical structure so that other downstream agents can deal with them.

However, some superoxides slip by and end up damaging important molecules and agents in your cell, including your DNA. Under certain conditions, the production of superoxides swamps the body's ability to denature them and then there will be significant oxidative stress to deal with. When a superoxide binds your DNA, it forms a lesion, often rendering that segment of your DNA less or completely nonfunctional. Luckily, your body also has enzymes that snip those lesions out and repairs the DNA segments. How do we know this? We can measure the number of repaired molecules in your urine, blood, and tissue samples and make a pretty accurate assessment of the level of oxidative stress in your body. However, oxidative stress does not occur equally across the entire body. Some organs seem to be more susceptible to oxidative stress, but assessing oxidative stress in each organ is often not practical and obviously fairly invasive since tissue samples are required.

So how is this relevant to aging in men? Men tend to have higher metabolic rates compared to women and therefore have the capacity to generate more oxidative stress throughout their lifetime, although there is significant variation depending on the biomarker of oxidative stress as well as other factors.[45] The differences in oxidative stress among various organs are small and vary between types of tissue, but over a lifetime, oxidative stress may contribute to the shorter life spans in men.

Other factors that contribute to oxidative stress are unhealthy behaviors such as smoking and alcohol consumption, which can dramatically increase levels of ROSs.[46] In other species, there is evidence that males exhibit greater oxidative stress for a number of reasons including the metabolism-promoting effects of testosterone and higher overall metabolism in males compared to females throughout their entire life course.[47] In fact, we are probably underestimating the impact of oxidative stress since all but a handful of studies have engaged relatively healthy, wealthy, Western industrialized populations. It is likely that much of the cost associated with oxidative stress is the deployment of oxidative stress defenses. To truly observe these costs, it would be useful to measure oxidative stress in populations that are not sedentary and well fed. Those that are energetically stressed, immunocompromised, and overall stretched to the limit in terms of life challenges probably endure greater costs associated with oxidative stress.

One of the few research groups looking at this question in a non-industrialized population is my own here at Yale. We are measuring two key indicators of oxidative stress among several populations including Shuar Native Americans in Ecuador and rural Polish women. Why these populations? First, they both experience greater degrees of energetic stress compared to sedentary Westernized populations, so the conditions are appropriate for observing evidence of life history trade-offs.[48] Our results thus far support the existence of a trade-off between reproductive effort and maintenance, at least in women.[49] Whether this also holds true for men remains to be seen. Overall, it appears that the slow, steady, and cumulative effects of higher metabolic rates and oxidative stress in men contrast sharply with the acute bursts of oxidative stress that accompany bouts of pregnancy and lactation in women. Men's oxidative stress costs therefore probably contribute to shorter life spans and greater morbidity than women.

DOES IT MATTER?

> Age and treachery will always triumph over youth and vigor.
>
> —Irven DeVore

The overall picture regarding physical fitness and age appears to be somewhat grim. Without the aid of sex and growth hormone supplementation, it is extremely difficult to turn back the hands of time. But does it matter? From an evolutionary perspective, humans have been very successful compared to most other primates and mammals. If one were to gather all of the other great apes in the world, chimpanzees, gorillas, and orangutans, this entire population would not be much larger than that of Yale's host city of New Haven, Connecticut. In contrast, there are over seven billion humans crowding our planet. Clearly, not having the physique of a twenty-year-old at older ages has not significantly affected our ability as a species to proliferate, thrive in a wide range of environments, and live well into our seventies, eighties, and sometimes nineties. Moreover, aging attenuates reproduction in men but does not prohibit it. Indeed, as we will see in a later chapter, older men have much greater capacity to reproduce than was initially suspected, even with diminishing muscles and growing waistlines.

As sexually dimorphic primates, human males and the males from our hominid ancestors almost certainly engaged in physical and non-physical forms of competition. Muscle and strength do make a difference, but this does not mean that males cannot adjust their reproductive strategies as their physical prowess starts to decline. Some fish change their sex in response to competitive and social factors, so it is not far-fetched for the males of our species to adjust their behaviors as muscle starts to decline and is replaced with fat. It is quite possible that having well-developed muscles at older ages became less important during our evolution. Indeed, there is compelling evidence in the fossil record showing that our overall morphology and physique have become more gracile compared to our immediate evolutionary relatives such as Homo erectus and earlier hominids.[50]

An interesting experiment would be to examine male fitness in contemporary hunter-gatherers and determine whether strength and vigor were correlated with important measures of success. During the course of my research among the Ache, I asked men to take a simple step test while I monitored their heart rate. Afterward I took salivary samples to measure their testosterone. I was attempting to determine whether they exhibited the same increases in testosterone in response to acute exercise as do men in industrialized societies. The results were inconclusive, but it was a good lesson in conducting field research.[51] Fast-forward a few years and Rob Walker of the University of Missouri and Kim Hill of Arizona State University replicated my crude study but with much more rigor. They conducted what I affectionately call the "Ache Olympics."

Walker and Hill asked Ache men and women to participate in several tasks that were meant to assess their overall physical strength and vigor. These included sprints, chin-ups, push-ups, and a number of other tasks. Not surprisingly, men were on average stronger than women. Both men and women in their twenties were the strongest and most physically fit.[52] Similar results have been replicated among other South American indigenous groups.[53] However, when Walker, Hill, and their collaborators

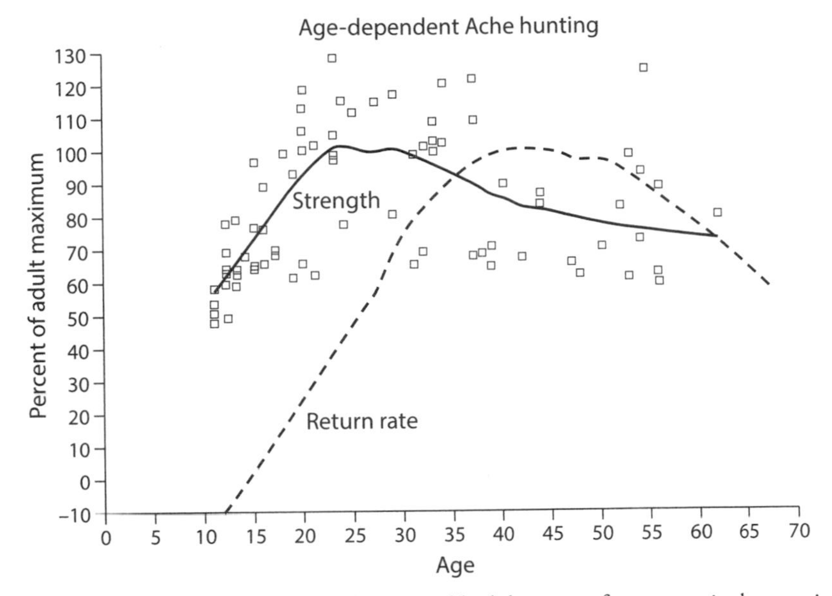

Figure 3.1. Hunting return rate, as determined by kilograms of meat acquired per unit of time, peaks two decades later than peak physical condition among Ache hunter/gatherer men.

Reprinted from R. Walker, K. Hill, H. Kaplan, and G. McMillan, "Age-Dependency in Hunting Ability among the Ache of Eastern Paraguay," *Journal of Human Evolution* 42, no. 6 (2002): 639–57 , copyright 2002, with permission from Elsevier.

assessed hunting returns and associated skills they found that older men had better game return rates than the younger ones.[54] In fact, men's return rate from hunting did not peak until their forties (Figure 3.1).

The implication is that for men, experience and wits are more important than raw strength, although physical strength is nonetheless vital for foraging success, as it is for other great apes. Any anthropologist who has attempted to keep up with men on a hunt can tell you it is physically demanding. However, compared to those of other animals, human male hunting returns are not as strongly associated with strength.

These results are intriguing since they dovetail with other human life history traits that are diagnostic and unique compared to those of other great apes and primates. Humans have large brains. In order to evolve these brains that account for about 20 percent of our basal metabolic rate and consume huge amounts of glucose, humans had to evolve the ability to secure high-quality and consistent food resources. A large brain requires a long life since it takes almost twenty years to grow a brain to full

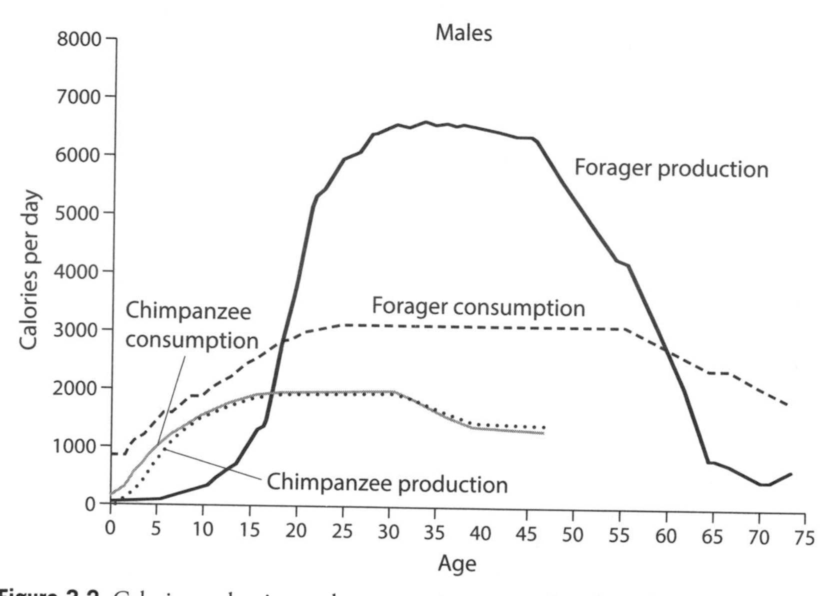

Figure 3.2. Caloric production and consumption averaged in three forager societies (Ache, Hadza, Hiwi) and wild chimpanzee populations.

H. Kaplan, K. Hill, J. Lancaster, and A. M. Hurtado "A Theory of Human Life History Evolution: Diet, Intelligence and Longevity," *Evolutionary Anthropology* 9 (2000):156–84. Courtesy of John Wiley and Sons.

size. For humans to attain a long life, extrinsic mortality and hazards in the environment had to be held at bay. Plus, what is the point of investing all this time and energy in a big brain if you don't have the time to fill it with useful information to justify its cost?

Several anthropologists have argued that male hunting provided crucial calories to support the evolution of many of the energetically expensive life history traits that are unique to humans (Figure 3.2).[55] If men could extend their time as effective hunters, drawing more on wits (brains) as opposed to muscle that is on the decline, this would give humans a significant advantage over other primates. When the foraging efficiency of male chimps is on the down slope, men are just hitting their stride. Given the amount of calories that men provide in addition to those provided by women in many foraging populations, this would be a huge advantage to anyone able to benefit from these resources.

Decreases in muscle mass and testosterone as well as increases in fat mass may undermine a man's ability to engage in physical competition, but it may open other opportunities for reproductive effort. Earlier I argued that while sexually dimorphic muscle tissue may represent investment in reproductive effort, fat deposition can be viewed as a way of augmenting survivorship. But what good is survivorship if reproductive opportunities are unavailable? This might seem like a good time to change strategies and address a need that evolved in our species: child care. If fat and decreases in muscle/testosterone could be leveraged to provide females with a resource that is sorely needed, then perhaps the door toward greater fitness at older ages is not closed. In fact, it may open a completely different range of possibilities. Paternal investment becomes possible.

CHAPTER 4

OLDER FATHERS, LONGER LIVES

> Amidst all the wonders recorded in holy writ no instance can be produced where a young woman from real inclination has preferred an old man.
>
> —George Washington to the Marquis de Lafayette, September 30, 1779

Seattle Seahawks football coach Pete Carroll wished he had more time during Super Bowl XLIX. Americans who wait until April 15 to file their income taxes wish they had more time. Authors on publication deadlines gladly accept more time. Additional time offers flexibility and the possibility of options to deal with life challenges. Having more time allows you to spread out energetic and resource costs and gives you the opportunity to rectify earlier deficiencies and errors. If one has the choice of servicing a debt, interest free, over a span of time or paying it all at once, it is often more palatable to choose the payment plan. We all wish we had more time.

Time is a resource that cannot be replenished or stored but is often an unavoidable constraint when facing immediate and long-term environmental and social challenges. Immediate challenges commonly include decisions and strategies on how one spends the day. One cannot be in two places at once, so allocation decisions are part of every organism's daily life. Those who make such decisions more efficiently than other individuals are likely to have a fitness advantage. An individual can look for food or mates, rest under a tree, or do any number of things. But it is difficult to engage in many activities at once. Multitasking is difficult and often results in one task receiving less attention and being performed with less attention and care compared to other tasks. The idea is that time spent in one task is at the cost of another. In evolutionary anthropology this is the primary area of interest in the field of human behavioral ecology.

Long-term challenges include the timing of life history events such as growth and reproduction. The total amount of time a species commonly has to grow, mature, and reproduce is reflected by that species' life span. The life span of humans is unique compared to those of other great apes. In short, human life span is significantly longer than one would theoretically expect.[1] How this evolved is still somewhat of a mystery, but some key ideas have emerged that make the scientific waters a bit less murky including the possibility that older men not only were the beneficiaries of the evolution of long life spans but also contributed to it. In essence, humans evolved the capacity to live several decades longer than our great ape cousins at least partially because of the evolution of novel reproductive strategies by older men. There are other compelling hypotheses that address human life span, but older men have seldom been considered as instrumental in the development/evolution of life spans.

Several individuals have lived more than 110 years. However, this is very rare. The probability of becoming a centenarian in the United States is pretty low and is influenced by factors such as sex, occupation, and race/ethnicity.[2] In other countries such as Japan, the probability is a bit higher.[3] Interestingly, women are much more likely to become centenarians than men, which makes sense given the other sex-specific mortality differences across the life span. Although the probability of living to the age of one hundred is higher than ever, most people will not reach this milestone. In the end, longevity records are not representative of the fates of most people and are of limited utility in assessing life span.

Referring to our U-shaped mortality curve (Figure 2.1), what is usually described on a graph with probability of death on the y-axis and age on the x-axis is a reflection of high mortality early and late in life. Focusing on the right side of the function, that is, changes in mortality with increasing age, one can see a population or species perspective of life span. As the right side of the U starts to rise, one can make the assumption that age-related mortality is becoming more prominent independent of environment. In the end, as the right side of the U starts to become vertical, it is clear that age-related sources of mortality are becoming the predominant factor, indicating an "expiration date" to the shelf life for that species.

There are important differences between classes of organisms such as reptiles, birds, fish, and mammals, so evolutionary ancestry and divergence need to be considered when making comparisons between species whose last common ancestor was deep in the evolutionary past. For example, reptiles tend to have longer life spans compared to mammals after controlling for associated traits such as body size.[4] There are many ways to use these data to define life span of a species. Some of these methods are quite involved and require familiarity with calculus and advanced mathematics. In the end, one can simply look at when the right side of the U graph starts to approach vertical. When this happens, essentially the probability of death is virtually 100 percent and you are more or less defining the common age at which death is likely due to aging and age-related illnesses.

Why is living longer advantageous? This may seem like a ridiculous question, but many other short-lived species have been very successful from an evolutionary perspective. Consider that beetles make up about three-quarters of the species on the planet and account for a good chunk of the earth's biomass. Rodents have also been extremely successful mammals despite their short life spans. What is the big deal about long lives?

Besides the obvious advantages of enjoying more years with family, friends, and the other wonderful things in life, there are many evolutionary advantages to living longer lives. With more time, an organism can grow more slowly and spread out associated costs. Human childhood is a good example. Between the ages of about five and eleven, children grow at a pretty steady rate. Energetic investment in childhood growth is slow and spread out over many years.[5] The only way this can be achieved is if extrinsic mortality risks are held at bay and children are allowed the time to grow slowly and steadily. Over a lifetime, organisms that can extend life span are able to spread out growth costs over a longer period of time. There are other benefits to long lives, slow growth, and large bodies, which we've covered earlier.

Body size and slow growth are often associated with decreases in extrinsic mortality. That is, when environmental hazards that contribute to mortality are kept in check, organisms can defer the costs of reproduction, grow larger, decrease metabolic costs, and possibly slow down the effects of aging. So how do organisms decrease extrinsic mortality? Some are just lucky. For example, on the island of Madagascar, there are species of primates that evolved for millions of years with very few predators and environmental hazards. These lemurs range from rat size to large raccoons. Yet for their body size and despite a highly variable climate, they live extraordinarily long lives. Imagine if your house cat lived to the age of thirty.[6] Similar patterns are seen in other small-bodied mammals that have lower extrinsic mortality such as naked mole rats, lorises, and bats.

But not all organisms were fortunate enough to evolve on isolated islands or, as in the case of naked mole rats, live in subterranean tunnels in the middle of the Kalahari Desert away from pretty much any environmental risk or predator. Some organisms evolved extended life spans by decreasing extrinsic mortality using social behavior. Primates are exceptionally good at this; this makes us unique compared to most other mammals. Most organisms reap the benefits of larger body size by evolving the means to grow faster. Primates as an entire order, on the other hand, grow *longer*, which means extended life spans allowed natural selection to select for larger body size, benefits and all.[7] As primates, humans are no exception.[8] Before humans evolved longer life spans, their evolutionary ancestors already had long life spans compared to other mammals, so in some sense humans benefited from an evolutionary head start as a result of the evolution of longevity in all primates.

Dolphins and larger whales give us a run for our money, but in terms of land mammals, primates are a unique evolutionary group. The idea is that social behaviors such as cooperation, group predator vigilance, and, in some cases, group foraging protect individuals from environmental hazards, allowing them to defer certain costs such as reproduction, spread out metabolic costs over a longer period of time, and in general select for a physiology that is more buffered against the detrimental effects of aging.

For social behavior and cooperation to evolve, organisms need to live together in relative peace and be able to track social relationships, recognize individuals, and have the overall brain capacity to keep the accounting of favors, alliances, and cheating behavior clear and accurate. This requires significant cognitive abilities, which makes it no surprise that long-lived, social animals also tend to have sophisticated brains and cognition.

Primates have evolved all of these traits and are among the most social animals on the planet.[9] This is especially true for males since male-male alliances are extremely important in human and chimpanzee societies. Although it is common for males in most primate groups to leave their natal group, only chimpanzees and humans exhibit the alternative great ape pattern with females emigrating and males staying in their natal group. Obviously humans have other social arrangements, but chimpanzees are certainly unique in this respect compared to other primates. It is therefore likely that the evolution of sociality in great apes, particularly humans and chimpanzees, involved a significant degree of influence from male-male social interactions.[10]

BUYING TIME

Extending life span often requires access to food resources that will support a long life. Organisms also need to evolve the ability to get through lean times. Fortunately (or unfortunately if you are struggling with weight gain), energy from food can be stored for later use. However, unlike energy, time cannot be stored. But one can "buy" time using energetic resources. The most salient example is fat tissue. During periods of environmental stress, fat can save the day, both literally and figuratively. The primary purpose of fat is to provide organisms with the flexibility to get through periods of energetic deficiencies. So if fat is such a great resource, why hasn't natural selection resulted in humans being super-fatty? In essence, it has—kind of. The current obesity epidemic speaks to our ability to sequester fat when there is an abundance of food. However, this may be an evolutionary novelty since having several thousand calories arrive at your doorstep with the mere pressing of a few buttons on a cell phone was not common during our evolution.

The efficiency with which our bodies store fat increases with age, which I'm sure you'll agree is a wonderful thing. Yes, I'm being sarcastic. Both men and women lose lean body mass and retain more fat as they get older. For both men and women, storing more fat may increase survivorship at later ages, but it can also come with costs related to cardiovascular health. It is unlikely that an increase in adiposity with age was the result of natural selection since fitness at older ages is limited, especially for women over the age of fifty when the probability of reproducing is pretty much nonexistent. However, it is possible that human males have leveraged this aspect of aging to their advantage. If adiposity increases adult survival, and the survival of older men and women is associated with the survival of their children or grandchildren, then there may be some selection to favor extended life expectancy.

But what is the point of surviving, whether with the help of more fat or not, if you cannot reproduce? For women, menopause makes reproduction impossible. However, men do not undergo menopause. Before we can entertain the possibility of older fathers contributing to the extension of human life spans, we need to establish men's capacity to produce any children to bounce on their knees.

OUT FOR THE SPERM COUNT

What do Robert DeNiro, Charlie Chaplin, Tony Randall, Michael Douglas, and Rod Stewart have in common? They all fathered children after the age of fifty. In Chaplin's case, he was seventy-three. Randall was seventy-eight. Fathering children at these advanced ages is less common than having children earlier in life, but it does happen at fairly high rates. Looking at fertility patterns of two representative industrialized populations, Japan and Germany, it is clear that peak fertility for men is in the late twenties (Figure 4.1).[11]

Men maintain the capacity to have children at older ages. The interesting questions are, then, why does fertility decline and why do some older men continue to reproduce and others do not? Clearly many males retain the capacity to reproduce later in life; however, most don't. Perhaps there is some degree of truth in all of the pharmaceutical advertisements that promote the use of virility-enhancing drugs. Or perhaps older males were not presented with the same opportunities to father children when they were younger as when they are older. Three possible sources of variation in male fertility emerge. They are (1) the ability to produce a sufficient number of viable sperm to fertilize an ovum; (2) the ability to engage in sexual intercourse; and (3) the ability to attract and procure mating opportunities with women of reproductive age. We will address each of these issues individually. But first we need to understand and define male fertility.

Most demographic studies are female focused since assessing a woman's fertility is fairly straightforward. While it can be a bit challenging to count the number of pregnancies since many conceptions are lost before

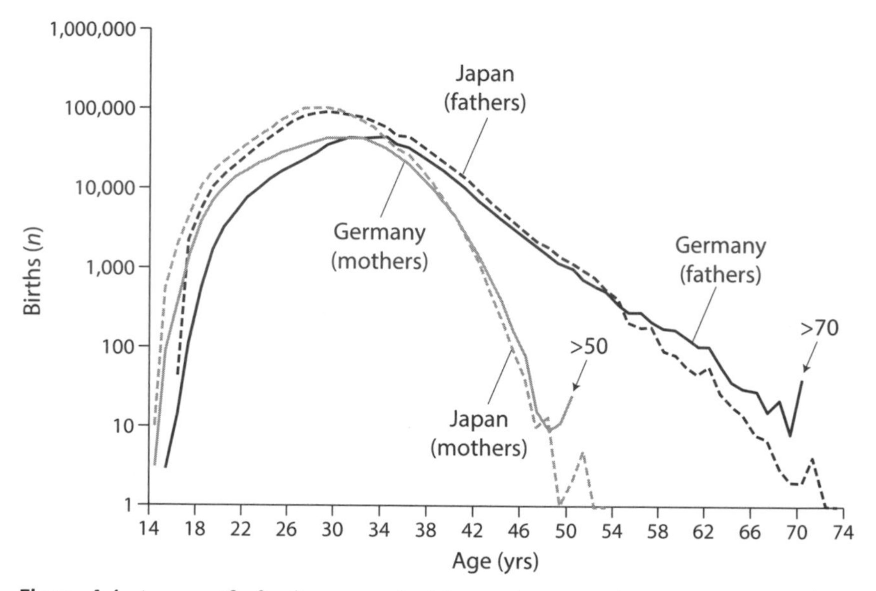

Figure 4.1. Age-specific fertility across the life span in men and women in two modern, industrialized populations.

B. Kuhnert and E. Nieschlag, "Reproductive Functions of the Ageing Male," *Human Reproduction Update* 10, no. 4 (2004): 327–39. Courtesy of Oxford University Press.

a woman even knows she is pregnant,[12] it is easy to ask a woman how many children she has had. For men, the question is not so simple, although one can attempt to assess male fertility at the population and individual levels.

Demographic analysis allows researchers to determine age-associated changes in fertility among males by examining population records of paternity and child survivorship, although there are some caveats that are specific to male analysis, the most important of which is caused by our old friend paternity uncertainty. That is, while demographic analysis of female fertility relies on accurate measurements of maternity, male-focused demographic analysis is a bit muddied by the fact that males may not always correctly identify their offspring. Depending on the population, the degree of paternal misidentification can range from 1 to 15 percent although the quality of the data also varies widely.[13] The development of internal fertilization has been and continues to be a powerful factor in the evolution of male reproductive strategies. While the risk of paternity uncertainty can vary significantly, one needs to assume a greater degree of error in assessing and estimating paternity compared to maternity.

From a physiological perspective, fertility is commonly assessed by sperm count and quality, that is, the ability and vigor of sperm mobility, the morphology of sperm, and the viability of genetic material. However, the utility of these measures in assessing male fertility, even in younger men, is less than robust. Using these factors to estimate the ability of a man to father children is fraught with uncertainty. Yet these measures are commonly used since other means of assessment have not been forthcoming. Is there some diminishment of sperm quality with age? Yes.

It is somewhat tricky to assess male fertility since the female partner's ability to conceive, non-sperm variables such as coital frequency, and erectile dysfunction can all affect a male's ability to father children. One way to focus strictly on sperm function and age is to assess the ability of the sperm of older men to fertilize an ovum without the confounds of sexual activity or coital frequency. There are several studies that have used this research method, but one of the most compelling was done by French researchers who looked at the statistical "risk" of conception during in vitro fertilization (IVF) trials in couples with proven female fertility and no male fertility factors. Here the sperm is introduced into a petri dish of sorts that contains the harvested egg of the female partner. The only variable was the age of both partners. What they found was an age-related effect on the part of males but not as much as what was expected.

When both partners were over the age of forty, the probability of conception failure was over five times higher than it was for couples who were in their twenties. However, when only the male partner was over fifty and the female partner was in her twenties, the risk of conception failure was only modestly higher and somewhat inconsistent. In essence, the study showed that while male age can be a factor in sperm quality, at least as assessed by conception success during IVF, the effect of male age was somewhat modest. By far the most important age-related effect was on the part of females.[14] The biochemical differences in the effect of age on gamete quality (ova and sperm) between females and males are somewhat unclear, but it is quite likely that the manner in which ova and sperm are produced is a factor.

All ova are produced in females during the second trimester of the fetal stage. After a series of initial cell divisions (meiosis) to halve the number of chromosomes in each cell, the ova go into a state of biochemical hibernation until puberty when the female reproductive system goes online and first ovulation occurs. In total, over a million primordial ova are produced, constituting the entire supply of ova that will be available for conception. Therefore, genetic anomalies accumulate simply by the nature of aging. There is some evidence of repair mechanisms that can attenuate some damage, but it is pretty clear that ova have a finite life span. In males, however, gamete (sperm) production is continuous, even during infancy and childhood, although production really gains steam during puberty. This constant production creates many defective or less than perfect sperm, but there are almost always a few thousand (if not millions of) viable sperm that are capable of fertilizing an ovum. Sperm supplies are constantly circulated, and unless there is a severe defect in the production mechanism, male fertility is more buffered against the effects of aging compared to female fertility.

The genetic contents of sperm have also garnered attention as factors in explaining human longevity. In males, attention has focused on the Y chromosome. Indeed, biologists have hypothesized that the lack of a second X chromosome may result in the expression of deleterious X-linked genes in males that are masked by the second X chromosome in females. The genetic studies of this possibility are ongoing. Moreover, non-genetic biomarkers have been sought, specifically a possible link between mortality and male fertility. As mentioned previously, an interesting negative association has been reported between the number of children a woman has and her postmenopausal life span.[15] However, a recent clinical research article has reported that male *in*fertility may be associated with greater mortality.

Does sperm count reflect other aspects of health and longevity? Male infertility is a problem for many men and couples; however, a recent investigation has reported that the implications may be more dire than originally thought. Men diagnosed with infertility had more than double the risk of death within seven years of diagnosis compared to men drawn from the general population.[16] Researchers compared mortality between men diagnosed with infertility at two separate infertility clinics, one in California, the other in Texas, and controls drawn from a social security national database. The sample size is very robust and the results are compelling. In summary, men with at least two indicators of male infertility, such as low sperm count and low semen volume, had 2.3 times greater risk of dying within seven years of being diagnosed.

Does this mean that men with fertility issues are condemned to shorter lives? Not necessarily. A knee-jerk response to the data might be that male infertility is a reflection of a deeper genetic imperfection that manifests itself in shorter life spans. Perhaps, but much more compelling follow-up research is necessary before we can arrive at that extreme conclusion. The authors of this investigation readily acknowledge that their results have limitations. For example, there is a lack of granular background information that would be very helpful, such as health history, lifestyle, occupation, and current overall health. The average age of the subjects was 36.6 and the average follow-up time period was 7.7 years, so while this study does not tell us much about older men, it does provide information on mortality risk.

Over the past couple of decades, the field of reproductive medicine has grown tremendously; this has been due primarily to advances in treatments and diagnoses. However, we also now have confirmation that health challenges associated with our Westernized lifestyle can influence reproductive function. For example, obesity can affect reproductive function in both men and women. In women, obesity and metabolic syndrome contribute to polycystic ovarian syndrome (PCOS) and a number of other hormonal issues. In men, obesity decreases testosterone levels and can lower sperm counts.[17] Similarly, many other illnesses including infections can compromise fertility.

Despite evidence that spermatogenesis is not as robust in older men compared to younger ones, the ability to maintain spermatogenesis preserves the possibility of having children at older ages. Spermatogenesis in older men is not perfect, but it seems to ramble on despite various challenges. With continued spermatogenesis and fertility at older ages, the possibility of additional fitness at older ages becomes very real. The implication is that traits that are associated with greater longevity can be passed through males. Coupled with the ability to obtain mates at these advanced ages, older men may have facilitated a central life history trait in humans: extended life spans.

FATHER TIME

Some biologists have argued that life span in humans is not so unique and that living past the age of last reproduction is not that unusual. There is some truth in that statement, but it is clear that while other organisms exhibit some extended life span beyond last reproduction, humans are alone in that about one-fourth to a third of female life is postmenopausal.[18] As mentioned previously, most mammals die around the same time they cease reproducing. In the case of humans, females live for two to three decades beyond the common age of reproductive cessation, that is, menopause, which occurs around the age of fifty. This is pretty much true for all human populations regardless of lifestyle, population, or genetic history, although some interesting patterns of variation are emerging.

How did this evolve? Several basic ideas have emerged that address this question. One is that humans probably did not live past the age of fifty in our evolutionary past. Perhaps longevity is a recent phenomenon due to the emergence of agriculture, urbanization, and other social safety nets. Evidence on longevity and mortality patterns in contemporary hunter-gatherer societies does not support this idea. They commonly have a life span that extends beyond menopause by roughly the same amount of time as that experienced in other populations. Yes, extrinsic mortality is higher, but the U-shaped curve in hunter-gatherer societies is pretty similar to that of more urbanized populations.[19]

Another idea is that the physical rigors of gestation and childbirth require women to cease reproduction in order to increase their survivorship to raise their children. In other words, as women age, their bodies' ability to withstand the physiological stresses of reproduction becomes ever more compromised, thereby becoming a hazard to their very survival. They therefore stop reproducing in order to be around to raise their final children, the idea being that, since human offspring are quite helpless and altricial, requiring tremendous amounts of care, it would be evolutionarily pointless to bear so many children only to have them die from lack of care because their mother succumbed trying to have more children. This argument hinges on two assumptions: (1) that there is the possibility of additional reproduction after the age of fifty and (2) that menopause is a physiological trait that can be adjusted. These assumptions are somewhat flawed since we know that menopause results from the depletion of viable ova. This is no mystery. All of the ova a woman has were created in utero and kept in arrested metaphase until menarche. As a woman ages, those eggs not only become depleted but also degrade in a fairly steady and predictable fashion. Menopause is a physiological

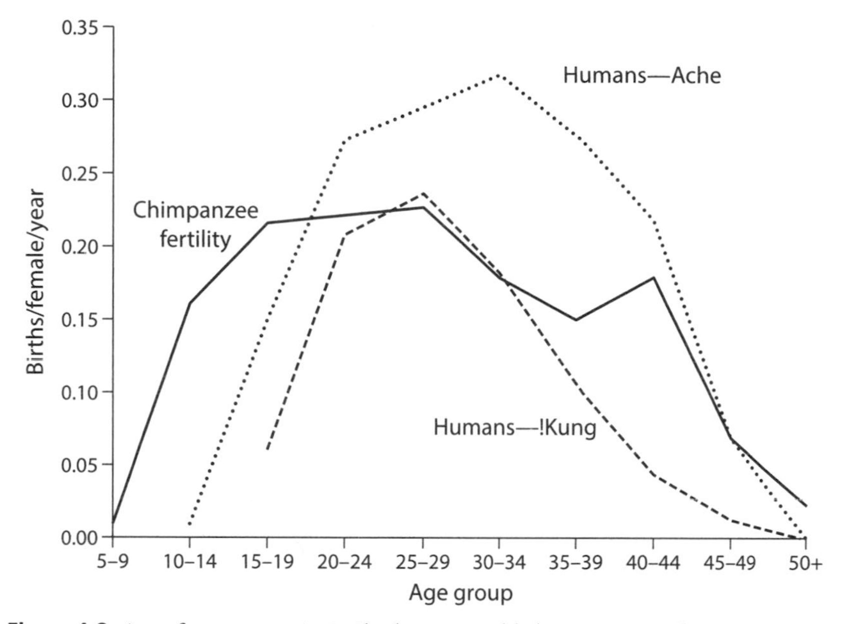

Figure 4.2. Age of menopause is similar between wild chimpanzees and representative forager populations. Note how reproduction begins earlier in chimpanzees but diminishes at about the same age as it does in the forager populations.

Reprinted from M. Emery Thompson et al., "Aging and Fertility Patterns in Wild Chimpanzees Provide Insights into the Evolution of Menopause," *Current Biology* 17, no. 24 (2007): 2150–56, copyright 2007, with permission from Elsevier. The Humans—Ache data come from K. Hill and A. M. Hurtado, *Ache Life History: The Ecology and Demography of a Foraging People* (Hawthorne, NY: Aldine de Gruyter, 1996).The Humans—!Kung data come from N. Howel, *Demography of the Dobe !Kung* (New York: Academic Press, 1979).

constraint that emerges from the biochemical limitations of ova viability. The same seems to be true for chimpanzees. Histological assessments of ovaries from chimpanzees that died of natural causes show almost identical age-associated patterns of depletion as humans.[20] Not surprisingly, age of menopause is about the same in chimpanzees, suggesting that our last common ancestor probably also ceased reproducing around the same age *and* that life span was tightly associated with reproductive cessation (Figure 4.2).[21]

However, the most commonly cited evolutionary explanation for menopause is the Grandmother Hypothesis proposed by Kristen Hawkes. Her idea has evolved over time, but her basic argument is that early in our evolution, prior to the emergence of extended life span, grandmothers who invested in the care and provisioning of their grandchildren enabled their daughters to have more children. Grandmothers shouldered some of the energetic and care burden of their daughters' children, thereby decreasing the energetic costs of reproduction in daughters and allowing them to have more children who had a better chance of surviving to adulthood and reproducing themselves. This hypothesis is somewhat rooted in the well-established finding that energetic expenditure and caloric intake are key determinants of female fertility.[22] As grandmothers assisted their daughters in raising their children, this promoted greater fertility in the daughters.

In essence, grandmothers were increasing their fitness through "inclusive fitness." What do we mean by "inclusive fitness"? In short, sexual reproduction results in the passing of half of a person's genes. This can be viewed as direct fitness. Inclusive fitness involves the passing of genes through individuals who share genes with you by common descent. By strict definition, increases in inclusive fitness must result from the direct actions of a related individual, such as a grandmother. If a grandmother helps with child care, resulting in an increase of her grandchild's chances of surviving and reproducing, then it can be said that the grandmother's inclusive fitness has increased. Grandmothering may increase the passing of genes through the greater rates of survival of grandchildren and through perhaps enabling an increase in the fertility of their daughters by alleviating some of their energetic burden. The thought is that grandmothers can provide significant additional value and resources to a hunter-gatherer family that can then be leveraged to increase their inclusive fitness.

The elegance behind Hawkes's idea is that it stemmed from her direct field research observations among Hadza foragers of Tanzania. It was evident that Hadza grandmothers provided assistance to the extended family and the community in general, not only through their food and labor contributions but by passing down life experience and knowledge to the younger generation. Clearly the presence of grandmothers was an evolutionary innovation that was important during human evolution. Grandmothers were important sources of allocare for children, a trait that is unique among great apes and mammals in general.

However, the fitness benefits provided by grandmothering are modest given the low degree of relatedness between grandmothers and their grandchildren. Nonetheless, recent mathematical modeling has shown that grandmothering could conceivably lead to extended life spans.[23] In essence the Grandmother Hypothesis provided a novel source of alloparenting and knowledge that can be passed across generations, but it does not explain menopause. In fact the physiology of menopause is quite simple. Ova seem to have a finite life span, most likely due to a biochemical constraint, and ultimately women run out of viable ova.[24] Why not produce more ova? Well, it does not matter. If all ova have a finite life span, it doesn't matter if you have one or a million ova. They will all degrade at the same pace and eventually become nonviable at around the same time.

Since ova accumulate a significant number of genetic defects, simply through aging, and a significant chunk of female life span is nonreproductive, the question transforms from "Why menopause?" to "Why such long lives?" Humans outlive their ova. So why do women have such a long life, particularly long postmenopausal lives? In addition to the Grandmother Hypothesis, other ideas have evolved that merit discussion, but they are beyond the scope of this book.[25] An answer may lie in male fertility at older ages.

To understand the possibility of a male contribution to human longevity, we need to understand some basic challenges regarding demography. For the past couple hundred years, demographers have grappled with the "two sex problem." Basically, in order to calculate and analyze population growth, demographers have focused almost exclusively on females for two reasons. As mentioned earlier, women can tell you with a fair amount of certainty how many children they've had. Again the challenge is paternity uncertainty. Therefore a demographer can calculate fertility at various age classes and across a woman's life span in a fairly straightforward manner. The same, however, is not true for men.

Because demographers have had only a modest degree of motivation to understand male fertility, it was generally assumed that male lifetime fertility tracked female fertility. After all, we are monogamous (right?), and once a woman's reproductive career had wrapped up with menopause, presumably a man's fertility would track this same trajectory. Therefore it was assumed, for good reason, that life span should coincide with the end of one's reproductive career since there would be no opportunity for natural selection to favor individuals at older ages. The eminent mathematician and actuary Benjamin Gompertz dubbed this the "wall of death" after the vertical line formed by the abrupt decline in male and female fertility and increase in mortality illustrated in demographic graphs. However, this conclusion, while sound, was based on the questionable assumption that all human populations followed the same pattern of age-related declines in male fertility as European men.

Biological anthropologist Frank Marlowe tested this assumption among the Hadza of Tanzania, the same forager group with whom Kristen Hawkes worked. Marlowe observed that although most men fathered children during their second and third decades of life, a significant number of men fathered children well into their fifties, sixties, and even seventies. (Paternity was self-reported and not supported independently by genetic paternity tests. However, assumptions of paternity seem to be reasonable, even if they are not infallible.) Marlowe's observations led him to suggest that perhaps demographers were missing the boat by assuming that male fertility tracked female fertility. Perhaps during our evolution many men continued to father children well after the age of fifty. This observation opened up a very interesting possibility that human longevity was being driven not by women but by men.[26]

Marlowe suggested that if men were selected to live longer, genes associated with longevity would be passed along not only to sons but also to daughters. In essence, females would be the beneficiaries of selection for longer lives in males. As female life spans became longer, their lives would outpace the viability of their ova, which are constrained to last only fifty years or so. He dubbed this the Patriarch Hypothesis.[27] It has always been assumed, based on Western models of demography, that male fertility generally tracks female fertility. That is, in a society that is thought to be monogamous, the number of offspring produced by a man will roughly be the same as the number of children his wife produces. This general model therefore assumes that the Western model of fertility is the most common among all human societies. This assumption leads to an erroneous conclusion; there is no reason why natural selection would favor longevity since there is no reproductive payoff in either men or women after the age of fifty or so.

However, the assumption of monogamy and a mirroring of male and female fertility is not entirely accurate. Most human societies, even today, are polygynous in some way or are characterized by a significant number of men older than fifty fathering children with younger, premenopausal women.[28] Therefore, it is possible for male fertility to extend well past the age of fifty if a man has the opportunity to have children with younger, premenopausal women. This is quite different from the experience of other great apes since male fertility coincides with longevity and physical condition. As chimpanzee males become old and feeble, mating opportunities fade. This contrasts with the experience of human male hunter-gatherers. Recall that in the previous chapter we saw how men's hunting returns, an important factor in attracting mates,[29] are highest long after their physical strength and fitness peak. Men are not as bound to physical condition for mating opportunities as are other great apes. This opened the opportunities and possibility for additional fitness for older men despite their waning physical condition.

Male fertility also extends well beyond the age of fifty since it is not as physiologically constrained by age as is female fertility. Spermatogenesis is not optimal at age sixty or seventy, but it is robust enough to maintain a significant capacity to father children.[30] Among forager populations, information on male fertility at older ages is limited, but among the Ache, for example, levels of gonadotropins, specifically follicle stimulating hormone (FSH), increase with age and are indicative of less capable but sufficient capacity to father children.[31] The question is, are premenopausal women having children by men who are older than fifty? If so, this would break down the "wall of death" and provide the fitness necessary to support and promote extended life spans. While the Grandmother Hypothesis relies on inclusive fitness to provide a mechanism for the evolution of extended life spans, a perspective that leans on direct male fitness would be very compelling.

Research by Stanford biologist and demographer Shripad Tuljapurkar and colleagues has shown that male fertility is quite significant in men older than fifty in many societies, both past and present. Again this is not to say that men older than fifty are producing offspring at the same rate as when they were in their twenties, thirties, or forties, but fertility is quite significant. Therefore, since these populations demonstrate significant fertility at older ages, natural selection would favor longer lives.[32]

As seen in Figure 4.3, male fertility in Canada tends to mirror the common assumption that male fertility at older ages is very limited. But when one looks at other populations, especially those that more accurately

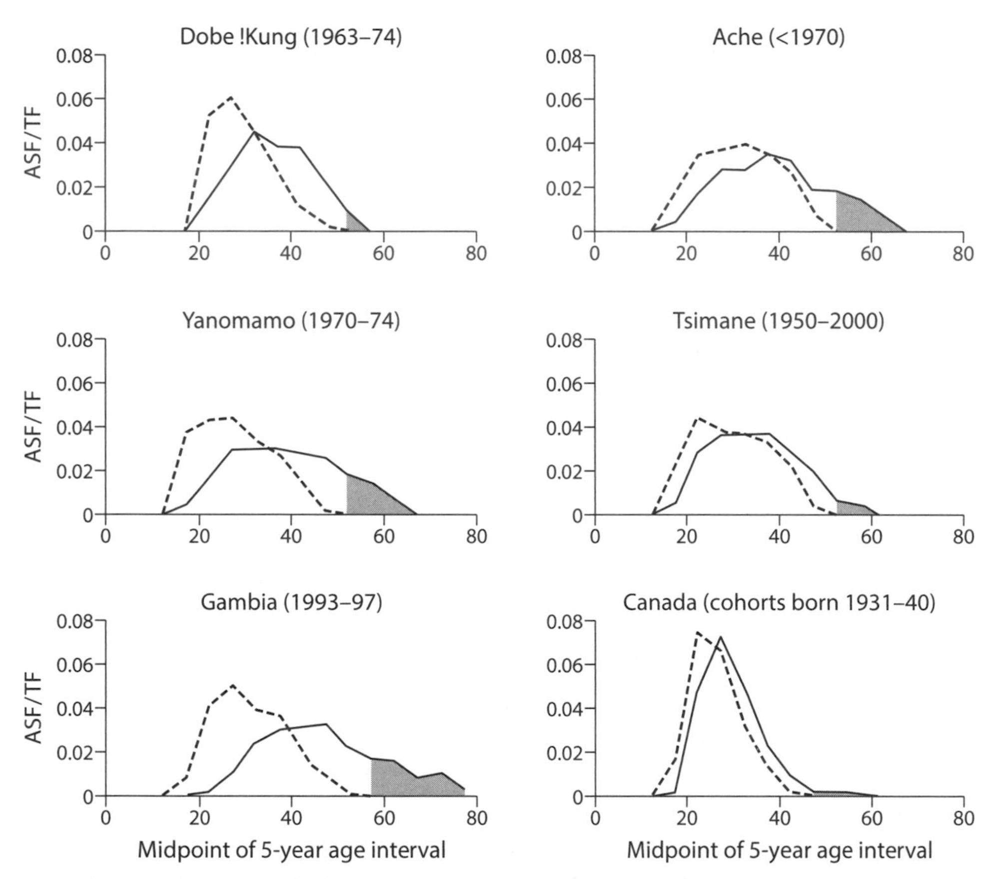

Figure 4.3. Age-specific fertility (ASF) as a percentage of total fertility (TF) illustrating male fertility (solid line) compared to female fertility (dotted line). Shaded area under male curve demonstrates additional male fertility after menopause. Note longer period of male fertility compared to females and significant amount of male fertility in non-Western populations compared to Western population (Canada).

S. D. Tuljapurkar, C. O. Puleston, and M. D. Gurven, "Why Men Matter: Mating Patterns Drive Evolution of Human Lifespan," *PLoS One* 2, no. 8 (2007): e785.

reflect the full range of male fertility variation around the world, it is evident that a significant number of men are still having children at older ages. This provides the fitness benefit that would contribute to favoring extended life spans in humans. This may help explain longevity in males, but what about females? It would seem that longevity genes that are positively selected for in males are likely to be passed on to daughters, even though there is no direct reproductive benefit to them. In essence, human longevity hypotheses that revolve around female postmenopausal fitness can be supplemented by hypotheses that engage with the demographic, biological, and behavioral evidence that male fitness at older ages has also contributed to the evolution of human life spans. Rigorous testing of male-centered approaches would be very challenging but certainly merited given the importance of this life history trait that distinguished us from other great apes and likely from our hominid ancestors.

But fathering children is not the same as being a father. Paternal investment in offspring and mates requires a unique set of traits that are evident in humans but not other great apes. In this chapter we have seen that older men have a much greater capacity for fathering offspring compared to other older male primates, mammals, and great apes. This may have contributed to the evolution of very long lives in humans. However, are there other opportunities for older men or even men in general to augment their fitness? Older men have the capacity to father children across the life span, but how do they engage with their offspring? Paternal investment presents an additional prospect for men to increase their fitness.

CHAPTER 5

DEAR OLD DAD

> To my deep mortification my father once said to me, "You care for nothing but shooting, dogs, and rat-catching, and you will be a disgrace to yourself and all your family." But my father, who was the kindest man I ever knew and whose memory I love with all my heart, must have been angry and somewhat unjust when he used such words.
>
> —Charles Darwin

Being a father can be complicated, which may be the reason I never became one. My wife and I know more than a few couples who, like us, do not have children. The reasons for not having children among these couples vary but, for the most part, we are all quite happy and lead fulfilling lives. However, every time I start working in a new field site, the locals ask if I have children. When I answer "no," they are without exception a bit shocked that I say this quite casually and with no remorse. Not having children can be devastating to many people, including men. Among the Ache and Shuar, the two societies among whom I have done most of my research, being a father is a defining feature of being a man. Fatherhood is important. However, the act of fathering children is different from bestowing paternal care. There is a significant amount of variability in the amount of care that is expressed by fathers. Nonetheless, the importance of paternal care is usually quite palpable in these societies. It is not uncommon for fathers to be engaged in the care of their children. Men can and do care for their children. This is a defining feature of our species compared to other great apes. How did this evolve and what role did older men play?

Anthropologists have written extensively about the importance of fatherhood in many societies. They write about the heartbreak felt by men who are unable to have children and the lengths to which they are willing to go to explore every option and chance to become fathers.[1] Biological anthropologists have also commented extensively on how men from numerous societies and subsistence lifestyles can be devoted fathers who invest tremendous amounts of time and energy in the care of their children.[2] However, sometimes a father can be a burden to a family, draining resources and inflicting verbal, emotional, and physical abuse on the mother and children.[3]

Fathering offspring is relatively easy. Being a caring father is not. In nature, males who keep in touch, both figuratively and literally, with their offspring and mates are rare, not just among great apes and other primates but among mammals as a whole. Males who invest in offspring through providing food and care or other tangible means are more common among birds but not mammals. The most likely explanation is that in species with external gestation, males can provide as much care as the females.

I'm not a bird, but I'll venture to say that sitting on an egg requires patience but not much skill. Male birds can sit on eggs just as well as females. Mammals, however, are constrained by internal gestation, which means that females bear the brunt of metabolic investment during pregnancy and afterward with lactation. Mammalian males cannot be pregnant.[4] Paternal investment is also often influenced by how a species reproduces. In species that reproduce by internal fertilization, male care is very uncommon since paternity uncertainty becomes a factor in the calculus of whether a male will invest in offspring. Some vertebrates, such as certain fish species, reproduce by external fertilization: females lay eggs in a nest and males deposit sperm as they swim by. This, along with guarding of the nest from other males who might also try to fertilize eggs, provides some assurance that the offspring belong to the depositor. Not surprisingly, the males of these species exhibit more paternal care.[5] So an important question to consider is: Under what circumstances would paternal care evolve? Perhaps the relationship with potential mates may have been important. Humans also engage in what is commonly called "pair bonding," that is, a male and female enter into an exclusive relationship that often results in paternal investment.

In this chapter, I propose that older men evolved the ability to leverage the physical effects of aging to engage in pair bonding and paternal care. The physiological changes that occur as a result of aging make them less physically capable of competing with younger men. However, because physical condition has taken on less importance over the course of human evolution compared to that of other great apes, older men have evolved the ability to increase fitness through paternal care and pair bonding. Indeed, the hormonal changes associated with male aging may actually prime older men to invest in paternal care. As life span increased, opportunities expanded for older men to increase their fitness. Similar opportunities may have evolved in women, but those in men have not received much attention or analysis.[6] For older men, aging removed certain reproductive strategies but allowed them to leverage other behaviors that would enable additional fitness. But before we discuss the particulars of older men, we need to understand the evolution of pair bonding and paternal investment in general.

THE PAIR-BONDED MAN

> Never marry at all, Dorian. Men marry because they are tired, women, because they are curious: both are disappointed.
>
> —Oscar Wilde, *The Picture of Dorian Gray*

I've been happily married for over twenty-five years now. I'm fairly confident my wife has been also. Besides the happiness that is inherent to a healthy and mutually fulfilling relationship, there is another nice perk. Married men tend to exhibit lower mortality rates after the young adult mortality bump compared to unmarried men.[7] At this point we have to restrict our conversation to those who are recorded as "married" in official census records, although it is quite likely that the same applies to those who are pair bonded and unmarried. Indeed, the hormonal associations with marriage seem to apply whether one has a marriage license or not.[8] Unfortunately, I am not aware of any research at the time of this writing on same-sex unions, although I strongly suspect future researchers will find similar results.

Why is being pair bonded good for men? There are two likely explanations. First, pair-bonded men tend to engage in less risky behavior. This is well documented in life and automobile insurance policies. Actuarial data strongly indicate that married men, especially those who are older but not incapacitated by age, are less likely to die as a result of risky driving behavior. One of the requirements for being a pilot of the SR-71 reconnaissance plane during the height of the Cold War was to be married, the rationale being that an older married man would be less likely to make rash or risky decisions that would jeopardize the mission. Second, it is good for a male's health to have a committed relationship with someone who cares or at least has a vested interest in him. Pair-bonded men tend to seek medical care more often than single men. There is also a growing body of compelling evidence demonstrating that the hormonal and metabolic milieu in pair-bonded men may facilitate greater longevity.

So why did pair bonding evolve? This is a wonderfully perplexing question that engages how we understand the evolution of male and female reproductive strategies. Pair bonding is certainly not common among mammals, much less primates. Other than humans, only a handful of monkeys engage in behavior that could approach pair bonding. Pair bonding is evident in certain New World primates including owl monkeys (*Aotus sp.*), tamarins (*Saguinus sp.*), marmosets (*Callithrix sp.*), and Titi monkeys (*Callicebus sp.*). The males of these South American primates form single-mate relationships with females and provide a significant amount of care to offspring, most notably carrying them and allowing mothers to do other things like forage more efficiently for food.[9] While the evolution of this unique trait is unclear, what is all but certain is that pair bonding likely provided fitness benefits that outweighed the option of not being pair bonded. The ecology may have demanded that both parents provide care to ensure survival of offspring that were altritial.

The neuroendocrinology of pair bonding is an extremely fascinating topic, one that has spawned numerous books and articles focusing on both human and non-human animal models.[10] For our purposes I will simply mention a recent study among the Tsimane in which salivary oxytocin and testosterone were positively associated with the duration of a hunt and time spent away from home. For those who are not familiar with oxytocin, I like to call this the "warm and fuzzy" hormone. It is well documented to be associated with bonding, primarily between mothers and infants. In more recent studies, oxytocin has been correlated with romantic bonding in humans. With regard to the Tsimane study, it suggested that men underwent hormonal changes that were indicative of a reinforcement of pair bonding and paternal investment. Although some caution is warranted since the functional significance of salivary oxytocin measurements with neurological activity is uncertain, one can make the cautious assumption that hunting makes the heart grow fonder.[11]

Although pair bonding has the potential to allow males to bask in the glow of higher oxytocin levels, there had to be some mechanism by which paternity uncertainty was mitigated since time away leaves a male vulnerable to other considerable evolutionary costs. For mammalian males, including men, the potential costs include cuckoldry, that is, investment in offspring that are not theirs. From an evolutionary perspective, this is an especially egregious cost since males not only lose time and energy investing in unrelated offspring but also possibly contribute to the fitness of rival males. Another potential evolutionary cost is the loss of mating opportunities with other females. Caring for offspring and staying with a single mate reduces the amount of time and energy that could be devoted to other mating opportunities. While this is also true for females, the potential fitness costs are greater for males since additional mating opportunities can result in significantly higher fitness benefits.[12]

Given these costs, how would paternal investment evolve? Clearly these potential costs are formidable barriers to the evolution of paternal investment or it would be much more common in mammals. The benefits had to outweigh the costs or the costs had to be mitigated in some way. One likely evolutionary scenario is that when offspring survivorship is positively associated with paternal survivorship, it is more likely that paternal investment will evolve. In penguins, this is certainly the case. Male penguins incubate eggs on their feet. If the male were to die, the egg would freeze almost instantly. Similarly, in humans, there is compelling evidence of decreased chance of survivorship in offspring in association with the death of the father.[13]

Ache men are characterized by a broad range of paternal engagement. Some men are great fathers, others not so much. I would guess that individual personalities are tremendous factors in that some men have the

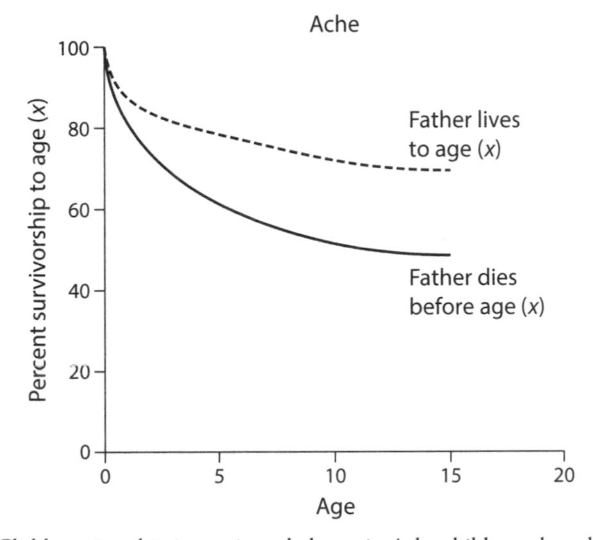

Figure 5.1. Child survivorship is consistently lower in Ache children when the father dies. Republished with permission of Aldine Transaction, from A. M. Hurtado and K. R. Hill, "Paternal Effect on Offspring Survivorship among Ache and Hiwi Hunter-Gatherers: Implications for Modeling Pair-Bond Stability," in *Father-Child Relations: Cultural and Biosocial Contexts Foundations of Human Behavior*, ed. B. S. Hewlett, Foundations of Human Behavior (Hawthorne, NY: Aldine de Gruyter, 1992), 31–56; permission conveyed through Copyright Clearance Center, Inc.

propensity to be affectionate and caring while others are more aloof toward their children. Nonetheless, Ache men provide care in the form of meat and other types of food. In my work among the Ache, I encountered children who were fatherless, either through death or abandonment; they were often emaciated or suffered from a lack of resources in other ways. It was very difficult for mothers to make a go of it alone. For example, children who had fathers exhibited much lower mortality than those who did not (Figure 5.1).[14] What is in it for men? At least among the Ache there are potential fitness benefits to being a good hunter. Those ranked as good hunters were more often identified as being the primary father,[15] and as we saw earlier, older men were often more efficient hunters than younger men.

It is therefore conceivable that as male life span increased, older males who invested in paternal care benefited by having greater offspring survivorship. It also speaks to the potential fitness benefits overall to investing in paternal care of offspring that are altritial and require huge amounts of care and resources. This would certainly present fitness opportunities for older men especially if the younger guys are deploying other reproductive strategies such as greater mate acquisition. However, this only shows the association between the absence or presence of a father on offspring survivorship. Paternal investment is about engagement with offspring and mates. How is this facilitated?

In the rare mammalian species that exhibit paternal investment, males display shifts in what can be considered reproductive states. In mammalian females, the three primary reproductive states are non-pregnant, pregnant, and lactating. It is very well established that increases in various reproductive hormones such as estrogens, prolactin, and oxytocin facilitate women's nurturing behavior toward their offspring.[16] In males, reproductive states are not as distinct, but they are evident nonetheless. That is, hormone profiles change to support paternal behavior. The classic example of this can be found in birds. In birds, hormones such as testosterone that are associated with mate-seeking behavior decrease, and hormones that facilitate bonding such as oxytocin increase.[17] Interestingly, some rodents exhibit the opposite effect: paternal behavior is supported by higher testosterone levels.[18] In humans, paternal investment and pair bonding are associated with lower testosterone.

When men become fathers, pair bond, or both, their reproductive hormone profiles tend to change. Testosterone in particular tends to decrease under these two circumstances compared to the time period during which they were non-fathers. Anthropologist Peter Gray has demonstrated the robust and cross-cultural nature of this effect in a number of populations including those in east Africa, China, and Jamaica.[19] The nature of the effect has also been shown longitudinally in studies examining testosterone levels in the same men before and after becoming fathers.[20] However, paternal behavior involves more than just fathering children. Engagement with offspring appears to be necessary to be able to observe decreases in testosterone.

Anthropologist Martin Muller and colleagues published an interesting study in which they compared testosterone levels between two culturally distinct Tanzanian communities living adjacent to each other. Their shared locality helped control for ecological variation that could affect things like nutrition and hormone values. One group, the Hadza, live primarily as hunter-gatherers, subsisting on wild game and plant products. The other group, the Datoga, are cattle herders. However, their differences extend beyond how they make a living. Hadza men spend considerably more time with their children compared to the Datoga. Hadza men hold, care for, and generally pay more attention to their children whereas Datoga men tend to be more detached and less involved in child-rearing.[21]

No differences were evident between the two populations when general measures of health and robustness and testosterone levels were compared. However, when testosterone levels were measured in men when they were in contact with children, interesting contrasts emerged. Hadza men exhibited significantly lower testosterone levels when they were engaged in the care of children compared to situations in which children were not present. Among the Datoga there were no such changes in testosterone levels. There was also an inverse relationship between the age of the youngest child in a Hadza household and the relative percent decrease in a father's evening testosterone levels. Perhaps this effect on Hadza fathers was due to disruption of sleep patterns or something else. But it is clear that having children around and being engaged in their lives influenced changes in testosterone levels that promote care and bonding. Plus it is a reminder that gross differences in testosterone do not seem to tell us much, but how men respond to children is important.[22]

This is all fascinating, but so what? What is the functional significance of changes in testosterone levels in association with fatherhood?[23] In other words, how do decreased testosterone levels promote paternal investment? Why would natural selection favor a decrease in testosterone in association with child care? Several possibilities emerge. Perhaps decreases in testosterone lessen a man's motivation to seek other mates and promote investment in children. But why would this strategy be preferable over seeking mates? One possible answer is that, for whatever reason, the male is not competitive with other males, either because of some perceived weakness or because of age. Another possibility takes an immunological angle. Perhaps decreased testosterone levels bolster a father's immune system and lower the risk of passing pathogens on to offspring.[24] This particular hypothesis is difficult to test since the addition of another person to an existing household with children introduces another potential vector. Therefore an appropriate test of this hypothesis would be to compare households of equal composition and size while varying

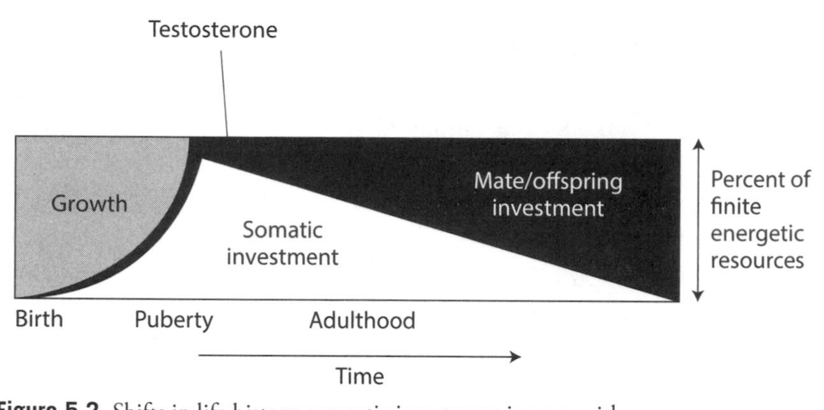

Figure 5.2. Shifts in life history energetic investment in men with age.
R. G. Bribiescas, "Testosterone as a Proximate Determinant of Somatic Energy Allocation in Human Males: Evidence from Ache Men of Eastern Paraguay" (PhD diss., Harvard University, 1997).

the paternity status of the adult male member of the family. Another test might be to observe the immunological response of fathers and non-fathers in circumstances where infection is high.

Regardless of the physiological mechanism, there should be an overarching hypothesis that would predict shifts in human male reproductive strategies with age. The previous chapters have provided evidence to support the argument that after the young adult bump in mortality, somatic changes with age in men instill a constraint that likely contributed to selection for a greater range of paternal investment strategies than what was commonly present in other primates, or for that matter other mammals. Figure 5.2 attempts to summarize this process by illustrating how men shift their investment focus from muscle and strength to paternal investment and greater investment in offspring and mates. Here I hope to provide some convergence of this evidence to show how these aspects of male aging biology lead to selection for traits such as paternal investment.

Because the level of testosterone declines with the onset of fatherhood in younger men, it is reasonable to assume that natural declines in testosterone levels with age might facilitate paternal investment in older men. Lower testosterone levels would mitigate mate-seeking behavior and favor the functional changes that would support paternal investment. In other words, the natural hormone changes with aging that alter the physical traits that would support greater fitness at older ages could be leveraged to support an alternative reproductive strategy: paternal investment. Another effect of lower testosterone levels is loss of muscle mass and increases in fat mass. This change in body composition not only causes men to shop for more comfortable trousers but also facilitates increased survivorship and, hypothetically, a hormonal milieu that would more effectively promote and support paternal investment. Perhaps in a modest way, men have evolved into the primate equivalent of penguins.

PUDGY DAD HYPOTHESIS

Older men have fewer options to increase their reproductive fitness. Typically an older guy simply cannot compete with younger males. We often don't have the muscle, vigor, or dashing good looks. We are also unlikely to have the neuroendocrine milieu to promote and support high-risk behavior that might be a bit attractive or lean body mass that often accompanies it. In many men, testosterone declines, fat reserves creep up, and the overall internal hormone environment is much more conducive to flab than to brawn. Many men try to keep up as they get older. They buy a new set of weights. Join a health club. Many take hormone supplements. Some men are quite dedicated and are successful at staying fit. Indeed, many older forager men will never set foot in a health club but are physically fit as a result of their daily activities and modest, lean diets. I don't know any man who has succeeded at rekindling the physique he had in his twenties. Some men come close but not without a lot of pharmaceutical help. My propensity to engage in high-risk behavior, even if it is just competitive sports, has declined since my teens and twenties. This doesn't mean I no longer watch football on Sunday afternoons or enjoy recreational physical activity. It just means that I am not the same man I was when I was younger. The fog of testosterone-induced physicality that enveloped me as a young man has lifted. Frankly, it's kind of nice.

As we've already discussed, older men tend to accumulate more fat tissue than they did when they were younger. Too much fat is often detrimental to health and well-being; however, fat is not inherently bad. Adipose cells that store fat, as well as the hormonal milieu that supports adiposity, evolved because it is important to have banked calories in the event of food shortage.[25] As our society and other parts of the world struggle with obesity and related metabolic and reproductive disorders,[26] we must remind ourselves that this was not the common condition during human evolution. Food shortage and insecurity were probably more common than caloric surplus and sedentism.

From a life history perspective, adiposity can be viewed as a signal and currency of investment in various traits such as reproduction and immune function.[27] Hormones that are associated with adiposity such as leptin and adiponectin act as messengers that inform the hypothalamus about fat storage status and often serve other purposes such as augmenting immune function. The deposition of fat as we age is likely a physiological constraint, one that is ingrained in the biology of our aging. But this does not mean that older men are without options. Indeed, there may have been selection for men to tolerate, and perhaps even promote, greater adiposity compared to younger men, especially if there is some benefit to putting on fat.

Chapter 3 discussed how men put on fat and lose muscle as they age. While this may be detrimental to attracting mates who are on the lookout for strong, muscular men, there may be an upside that men have exploited much more so compared to other great apes. It is common for male fertility to coincide with peak physical condition. However, male reproductive value tends to peak around the third decade of life. This seems a bit counterintuitive, suggesting that some other factor independent of strength and vigor is driving peak fertility.

For paternal investment to evolve, there should be some association between paternal and offspring survivorship. Therefore, one might predict that greater adiposity in older men is associated with paternal investment or is an alternate signal of attractiveness. A recent investigation showed that women preferred men with some degree of adiposity and that adiposity was associated with testosterone and enhanced immune function, suggesting that women were preferring men who had a good probability of surviving and perhaps providing some degree of paternal care.[28] Interestingly, comparing leptin levels, a hormonal reflection of adiposity across ages in chimpanzees and humans reveals that male chimpanzees seem to invest solely in lean tissue whereas men add on fat as they get older.[29] In other words, compared to humans, there has been little selection for older male chimpanzees to put on fat, perhaps signaling a lack of any evolutionary incentive to do so.

Here I suggest the Pudgy Dad Hypothesis: increased adiposity, including fat deposition that is facilitated by age, was leveraged during human evolution to increase paternal survivorship, mitigate mate-seeking behavior, and promote a hormonal milieu that supported paternal care. More specifically, decreases in testosterone in association with aging and greater adiposity contributed to attenuating mate seeking and promoted offspring and mate care. As discussed in our previous chapter, adipose tissue contains the enzyme aromatase, which converts testosterone into estradiol, the common female reproductive steroid that may also contribute to decreases in testosterone and behavioral shifts that are associated with offspring care.

How can we test this hypothesis? The obvious way is to conduct a controlled experiment in which we manipulate both hormone levels in men and exposure to their offspring. Clearly we cannot do that. For now we are left with naturalistic experiments that aim to provide some support for this hypothesis. A recent study using a database of over ten thousand men tracked various measures including body mass index (BMI) through the transition to fatherhood. They found that after controlling for relevant factors, men who became fathers exhibited a slight but significant increase in BMI. Men who were resident fathers exhibited the greatest increase.[30] This is modest evidence, but it is a start.

Clearly men can engage with their children through a broad range of behaviors. This plasticity of behavior toward their children is in itself unique and worthy of some discussion. There are no other primates that exhibit a range of behaviors. The question, then, is why did the broad range of engagement emerge and how did it evolve?

FLEX PLANS

My wife is a big believer in stretching. According to her one can resolve most kinks, aches, and possibly the national deficit with a good stretch. Indeed being human is all about flexibility and, more important, adaptability. In evolutionary biology terms, to be "adaptive" is to exhibit a trait that was shaped by natural selection. "Adaptability" is the expression of a trait that allows you to adjust your phenotype on the fly in response to environmental variability. An example of this is tanning. For some, exposure to the sun causes a darkening of the skin due to the responses of melanocytes. Some people are responsive to the sun, others not so much. Adaptability is the range of variation that can be expressed by a phenotype. In many ways, behavior is the ultimate expression of adaptability since it can be manifested across a broad range of possibilities.

The ability of our species to exhibit high fertility and produce offspring that have extended periods of growth and development requires tremendous amounts of care; this is an evolutionary novelty. It is difficult to imagine a scenario in which women would be able to have so many stacked children that required huge amounts of resources and care without some outside help. Some might assume that paternal care is something that should come "naturally." As television-watching Americans, many of us grew up with shows that highlighted the perfect father who doted over children with wit, humor, and love. We therefore think this is likely the norm. When one looks at a newborn, it is difficult not to be moved and restrain from pinching cheeks. Granted some men are skittish around babies and children, mostly out of fear of "breaking" them. It is a fragile life just begging for care. But there is a darker side to the relationship between children and males. Infanticide is not uncommon in the animal world. Humans are no exception. Among lions, males who take over a pride and run the alpha male out usually kill all the cubs. The evolutionary rationale is that as long as the lionesses are nursing those cubs, she will not go into estrous and will be unwilling to mate. Since the tenure of an alpha lion tends to be short, brutal, and uncertain, a new alpha does not have the luxury of waiting until the cubs are weaned to mate with the females. The neuroendocrinology of nursing and the resumption of ovarian function support this notion since suckling by cubs stimulates the production of hormones such as prolactin that suppress hypothalamic function and the resumption of ovarian function and ovulation. Infanticide is also fairly common in great apes and primates for similar reasons.[31]

Humans are not lions, so we cannot make direct comparisons between the two groups. Nonetheless, men are more likely to abuse or kill children of spouses/mates fathered by other men than they are their own children. Again, paternity uncertainty is likely to be a factor. When viewed in comparison with other primates and mammals, it is clear that men exhibit the greatest plasticity and flexibility with regard to engagement with their

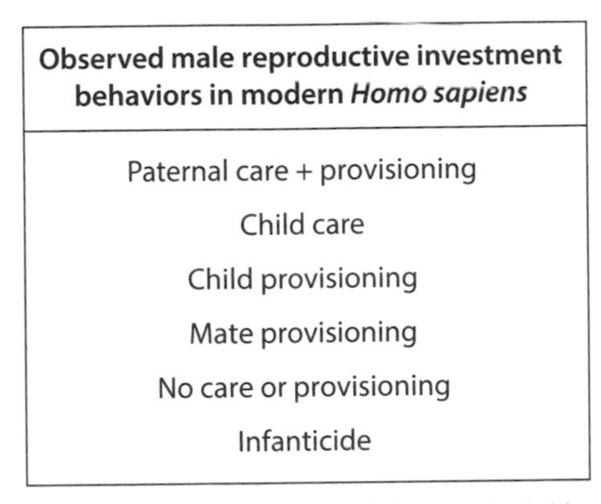

Figure 5.3. The range of engagement of human fathers with children.
R. G. Bribiescas, P. T. Ellison, and P. B. Gray, "Male Life History, Reproductive Effort, and the Evolution of the Genus Homo: New Directions and Perspectives," *Current Anthropology* 53, no. S6 (2012): S424–S435.

children. Men exhibit a broad range of care (or lack thereof) behaviors compared to other primates that exhibit paternal care (Figure 5.3).[32]

How and when do men shift or modulate their care and investment? Do all men have the potential to be infanticidal or caring? Most important, for the arguments in this book, does age play a role? However, before we address these questions, we have to ask why other primates do not exhibit the same range of behavioral plasticity. If keeping a suite of behavioral repertoires in your quiver grants the advantage of responding in the optimal way, why isn't behavioral plasticity more common?[33] The answer I would argue lies in the costs of plasticity. Adjusting your physiology can be metabolically costly since it usually involves breaking down and rebuilding tissue. For example, a deer builds muscle tissue in its neck in preparation for rutting, but by the end of this period of sexual activity, such tissue is broken down. This can be metabolically pricey. The metabolic costs for behavior are not the same, but there can be costs if there are wrong choices.[34] Evolving the ability to assess and make the optimal paternal investment choices seems to have been important for older men. The experience acquired through aging might be the mechanism through which those optimal choices are made.

I suggest that while younger men can certainly be good fathers, for physiological and behaviorally strategic reasons, older men are more likely than younger men to engage in paternal investment and gain a greater potential fitness benefit. In essence, the physical effects of aging will constrain reproductive strategy options available to men, leaving paternal investment as a particularly viable reproductive strategy for older men. As men aged, the need for behavioral flexibility became more apparent. With longer life spans, men were able to enter into a life stage that could take advantage of paternal investment. Older men became the plastic fathers.

In addition to the benefits to offspring, paternal investment may have provided a perk for older fathers. That is, as older men became more paternal, selection favored even longer life spans in both males and females. Men who have children tend to live longer lives than those who do not. Interestingly, though, the act of fathering children does not confer this longevity effect. A father needs to stick around and participate in raising the child in order to reap the longevity dividend. Perhaps there is no fatherhood benefit and what is observed is a non-fatherhood cost. Men who do not have or invest in children might behave or exhibit physiological processes that compromise their well-being. I suspect the answer probably lies somewhere in the middle.

There are interesting effects on male longevity depending on the sex of a man's children. Recall the Polish demographic data that showed that not only did men with children live longer compared to non-fathers but having daughters amplified this effect. The researchers suggested that daughters contribute to the care and well-being of their fathers in ways that sons do not, such as making sure they eat properly, taking them to the doctor, and overall simply paying attention to health hazards that sons often ignore, not only in their fathers but in themselves.[35] Why the sex discrepancy in caring for dear old dad? Cultural norms surely play a role, illustrating the important interplay between culture and biology in human evolution. It could also be that in some societies sons disperse in search of mates while daughters stay home waiting for potential suitors. Even after daughters marry and have children of their own, they are more likely to return home more often and tend to the well-being of their parents more so than sons.

MY DEAR OL' GRANDDAD

Some studies have suggested that grandmothers are able to augment their fitness through investment in grandchildren. The benefits might include directly increasing the survivorship of their grandkids or alleviating the burden on their daughters so they can have more children. Although it is certainly true that having grandmothers around can sometimes have a positive effect on grandchildren, the evidence indicating that this would account for the evolution of longevity in humans is still unfolding.[36]

What about grandfathers? The evolution of longevity in men is not as much of a mystery since, unlike women, men enjoy some degree of fitness and fertility after the age of fifty.[37] However, it can be hypothesized that older men, grandfathers in particular, could potentially add to their lifetime fitness through investing in grandchildren. In addition, grandfathers could potentially contribute to the evolution of other life history traits such as extended childhood by participating in the care and provisioning of grandchildren. In humans, the period of juvenility in which offspring are more or less dependent on care is very long compared to that of other primates and mammals.[38] It is unlikely that such a trait could evolve without the aid of extra-parental care.

A comparative examination across several different populations investigated whether there was any association between the presence of grandfathers and important factors such as grandchild mortality or sheer numbers of grandchildren as an estimate of direct fitness. Evolutionary demographer Mirkka Lahdenpera and colleagues examined historical demographic records from preindustrial Finland and eleven other populations from earlier studies. Their meta-analysis revealed very little evidence of any effect of grandfather presence on age at first reproduction, interbirth interval, reproductive tenure length, lifetime fecundity (i.e., number of born children), grandchild survival, lifetime reproductive success (i.e., number of children sired surviving to adulthood), or number of born grandchildren. There was also no association between the number of grandchildren and male longevity. This study was comprehensive, yet in the end, there was very little evidence that grandfathers make much of a difference for the number or survivorship of grandchildren.[39]

In some cases, the presence of a grandfather was actually associated with lower rates of child survivorship.[40] So much for granddads. But we must remember that the assumption of Lahdenpera and colleagues was that the relevance of grandfathers increased if monogamy was the common mating system during human evolution since the constraints found in women would also be the same for men. That is, in monogamous situations, when women stopped reproducing, so did the men. However, it is all but certain that men exhibit significant fertility after the age of fifty, either through polygyny or by taking on a younger mate.[41] The lack of a grandfather effect also supports the hypothesis that selection for long lives in men, and likely women, was the result of direct fitness benefits of older men fathering children, not through grandfathering.

This does not mean that grandfathers cannot or do not invest in grandchildren. Clearly different types of inheritance (social standing, titles, and resources) can be and are passed from one generation to the next. In humans the ability to transfer resources between generations is an extremely important consideration, one that has been proposed as having an influence on a number of life history traits, including the evolution of longevity.[42] Nonetheless, it is apparent that grandfathers do not have the same impact as grandmothers when it comes to investing in child care or provisioning. Why is this? Perhaps men become more frail as they age and simply cannot perform the same tasks as grandmothers. This may be the case, but one would assume that grandfathers are surely able to watch grandchildren as well as grandmothers. A more likely explanation again may be that paternity uncertainty is playing a role. An important prerequisite for the evolution of paternal investment is a fairly high degree of paternity certainty. In some indigenous South American groups, paternity is assigned in a more or less probabilistic manner. For example, among the Ache, a primary, secondary, and tertiary father is identified with each child, depending on how much each man "contributed" to the conception since it is not uncommon for women to have had multiple sex partners around the time of conception.[43] Among other Amazonian groups, paternity is partible. That is, paternity can be assigned across multiple men.[44]

After the age of fifty or so, men can father children *and* be grandfathers at the same time. This is quite unusual since this is impossible for postmenopausal women and very rare in other mammals. For this to occur, older men need to have children with younger, premenopausal women. Clearly from an evolutionary biology perspective, there are fitness advantages for older men to continue to reproduce. But are there any costs or constraints? There must be or one might predict that every man over the age of fifty would be having children. Obviously this is not the case.

So what are the constraints? For starters, older men would have to convince younger women to have sex with them. In many cultures, younger women often have a limited or no choice as to whom they take as a husband. Therefore, depending on the cultural circumstances, a man can either convince a woman to mate with him or negotiate with a third party, perhaps the woman's father, to allow her to marry him. We will leave aside the more disturbing possibility of rape since the evolutionary perspective on this horror is fraught with complications that are beyond the scope of this book and require a much more thorough, nuanced, and sensitive discussion.[45]

What is clear is that older men evolved the capacity to "stack" offspring across generations in the form of continuing to father children after they become grandfathers. This is not unique to men since women can be both mothers and grandmothers later in life, although they cannot produce new children postmenopausally. However, men have a greater opportunity to father additional children because of sustained spermatogenesis and fertility after the age of fifty. This would likely be associated with polygyny or serial mates since the mothers of a man's grandchildren would likely be menopausal. In societies in which polygyny is common, such stacking of offspring would represent an evolutionarily unique opportunity to increase fitness. According to the data from Tuljapurkar and colleagues (Figure 4.3), it appears that this may have happened with greater frequency than demographers have assumed. Moreover, testosterone data from polygynous men in Kenya show that when men are fathering children with multiple mates, their testosterone levels are higher than those of non-polygynous men.

Harkening back to the previous chapter, older men have evolved the ability to forgo physical strength and deploy political and economic skills to access resources that younger men often do not have (Figure 3.1). As a result, polygyny and paternal investment became a viable option to gain greater fitness at older ages. Being a grandfather may add polish to fitness, but being a father at grandfatherly ages is a clear and unique way to tack on extra fitness.

Until now, we've discussed how older men can contribute to their own fitness as well as to the evolution of important and defining life history traits. We now turn to more immediate issues related to men's health. After all, men cannot father children or contribute to human evolution if they are not in good health. Or can they?

CHAPTER 6

DARWINIAN HEALTH AND OTHER CONTRADICTIONS

> Evolution does not select for health.
>
> —Peter Ellison

The previous chapters have, in my humble opinion, provided a compelling argument for how male aging evolved and how older men might have influenced human evolution. As this book winds down, I would venture to guess that you might be saying, "Ok, this is all very interesting, but what's this got to do with me?" This is a fair question. Also, what are the primary health issues in older men? Why are they particular to older men? How did they evolve? Are they common across all human populations? What if anything can be done about them? There are many aspects of male health that elude a thorough understanding by the medical community. I would argue that evolutionary and life history theory could provide some insights into the emergence of certain diseases. But I am not the only one.

The burgeoning field of evolutionary medicine continues to shed new light on the role of natural selection and evolution on the development, incidence, and etiology of numerous diseases ranging from cancer to diabetes to mental illness. Evolutionary biologists and clinicians are engaged in a thriving conversation about how awareness, knowledge, and understanding of evolutionary and life history theory can contribute to understanding disease and illness. In 2009, the National Academy of Sciences brought together an unprecedented group of biologists and clinicians, including the deans of the nation's most prestigious medical schools all participating in impressive presentations on how various illnesses can be better understood through the lens of evolutionary and life history theory. Enthusiastic handshakes, pats on the back, and promises to reengage were the story of the day as scholars and scientists returned to their home institutions, invigorated with the motivation to set a new research agenda.[1] Progress has been a bit slow, but that is to be expected when new paradigms are introduced to curricula and research agendas. Nonetheless, it would be great if the National Institutes of Health could develop a grant program that focused on evolutionary medicine and human evolutionary biology. But I digress.

Caution is merited when trying to deploy an understanding of evolution to promote greater health. While evolutionary medicine provides powerful explanations for important health challenges such as antibiotic resistance, reproductive cancers, and metabolic syndrome, as well as having the potential to contribute to the development of new treatment and prevention strategies, it is vital to remember that natural selection does not favor traits that promote health unless there is a fitness benefit.[2] Moreover, it is often not possible to optimize multiple physiological functions owing to trade-offs that are inherent to our basic biology.[3] If I may take the liberty of personifying the driving force of evolution, natural selection does not care about your health or if you feel good. Over the years, I have heard many lectures and interviews on how we as a species, population, community, and so forth would be healthier if we simply understood and embraced our evolutionary past. Very bright scholars have espoused the belief that cancer and diabetes are products of a mismatch between how evolution "intended" us to live and how we are victimized by our current "unnatural" lifestyles of sedentism and high-calorie diets. There is more than a grain of truth in these perspectives, especially in light of reproductive cancers and diabetes. However, the engagement of evolutionary theory with our well-being is grounded in how we define health and illness.

Few topics garner more attention than diet and nutrition. What is the optimal diet for older men or humans in general? Bookstores devote entire sections to this question. In her book *The Fragile Wisdom: An Evolutionary View on Women's Biology and Health*, Grazyna Jasienska makes the compelling argument that healthy diets are somewhat of a myth. This is not to say that what you eat is unimportant, but the concept of the perfect, one-size-fits-all of humanity healthy diet is unrealistic. People's cuisine varies considerably across populations, and what might be good for one person might not have the same benefit for someone else simply because of common biological variation. The fact that there are seven billion of us on the planet strongly suggests that we are eating quite well at least as far as our fertility is concerned. Again, Jasienska is not giving the green light to binge on thick-sliced bacon, ice cream, and cheeseburgers. She simply suggests that a biologically meaningful definition of "health" is nuanced, multifaceted, and colored by our understanding of life history variables such as reproductive fitness, survivorship, and pleiotropic effects, which are, as we mentioned in chapter 2, the interaction of the expression of traits. There are few if any nutritional magic bullets.

For example, zinc has received a considerable amount of attention as it pertains to male health; the interest in this mineral has been due to research suggesting that it has an effect on male fertility.[4] The evidence is a bit scant, although it does seem that the total elimination of zinc from a male's diet will affect spermatogenesis. In a modern context, it would be extremely difficult to be zinc deficient by any measure; however, it is not uncommon in less developed areas of the world where diets can be simple and often unchanging. Even in these cases, there is no evidence to suggest that male fertility is compromised.

Jasienska's views are echoed in a book chapter that I coauthored with Peter Ellison a few years back at the request of Steve Stearns, professor of ecology and evolutionary biology at Yale and one of the leaders in life history theory and biology. In this coedited volume with Jacob Koella on evolutionary medicine,[5] Peter and I wrote about trade-offs associated with reproductive endocrinology in men and women. Peter discussed the female side of things and I took on the guys. In our chapter, we independently converged on the perspective that reproductive hormones such as estrogen and testosterone augmented reproductive "health" early in life but later in life could be considered "unhealthy," contributing to an increased risk for breast or prostate cancer.[6] This conversation continued after the publication of this edited volume.

At a conference symposium a few years ago on evolution and health, I suggested that if I were to discover a common, worldwide condition that was all but certain to shorten a woman's life, some very motivated people would set out to eradicate this hazard. Dare I say millions of dollars of grant money would be spent and foundations would emerge to take on this threat to women's health. This would be quite challenging since the condition is childbearing. Women who have more children are at greater risk of having shorter life spans, especially when women struggle to attain the necessary calories to support their immune function, oxidative stress defenses, and overall daily needs.[7]

Does this make pregnancy and lactation without complications a health hazard? It would be very awkward to make this argument. It is well-known that childbearing involves more risk for older women than for those in their twenties. But it is problematic to tell a woman in her twenties that the child in her womb will likely take a few years off her life, even if she adheres to the purest and most recommended diets and lifestyles. What is all but certain in evolutionary biology is that we are not designed for longevity. We are designed to live long enough to create the optimal number of copies of ourselves within the context of others who are trying to do the same. If evolutionary biology has selected a particular type of health, it is reproductive health. Longevity and vigor only serve to provide us the time to optimize our reproduction.

We can invoke this argument for men's health as they age. Testosterone supplementation does have very powerful effects. It decreases fat deposition, increases muscle mass, gives libido a bit of a boost, and causes men to feel better about themselves. A few years ago, I was quite skeptical about testosterone supplementation and still have serious reservations. However, it is difficult to ignore the positive reports from men who have taken the supplements, especially those who were genuinely hypogonadal. But as anyone who has had one martini too many can tell you, feeling good does not necessarily mean you are engaging in healthy behavior.

The effects of testosterone on longevity and aging are more nuanced than what is captured by simply stating that they are bad for long-term health. One research strategy that is commonly deployed by biologists is to observe the effects of adjusting the variable of interest toward one extreme or another. For example, remove a hormone altogether or crank up the dose to a very high level. Since it is sometimes unclear what merits being considered a "high" dose, it is often easier to simply remove the agent and see what happens. You cannot go lower than zero. In the case of testosterone, we should therefore be interested in the longevity and health of men who do not have any testosterone. One such condition is a genetic disorder known as idiopathic hypogonadotropic hypogonadism (IHH). Here the gene responsible for producing GnRH and kick-starting a male's (and female's) reproductive system is nonfunctional. No GnRH, no LH, and therefore no testosterone.[8] A wee bit of other androgens (testosterone-like hormones) may squeak into the system from the adrenal glands, but they are marginal.

The effects of IHH on male longevity are not yet known, but what is clear is that IHH is a much more heterogeneous condition than first suspected. Many gene loci are involved, and there is a significant gene-environment interaction.[9] In short, this model of complete testosterone deficiency has many moving parts and is less than perfect for understanding the potential long-term effects of having no testosterone on aging and life span. What we need to do is experimentally manipulate hormone levels in a controlled manner. It is time to talk about castration. I'll give the gentlemen a moment to compose themselves.

Castration, both surgical and pharmacological, has been a common form of experimentation in animals. Neurobiologist Robert Sapolsky noted that the era of behavioral endocrinology was ushered in when a farmer first lopped off the testicles of a surly bull.[10] He was right. Removing or inactivating the endocrine glands that produce testosterone often extends the life of the animal. Yes, some of those years are saved because fighting and injurious behavior are eliminated, but even if you control for the behavioral aspects of testosterone, male animals tend to live longer. It is obviously unethical to conduct such experiments in men. However, there are historical records of ritual castration in several societies. In late nineteenth-century China and the Ottoman Empire, men of certain religious sects underwent not only castration but complete genital removal including the penis and scrotum.[11] In preindustrial Korea, eunuchs were common in the courts of royalty as well as in boys' choirs in seventeenth-century Europe.[12] There are many other instances of castration and genital mutilation in the ethnographic record, but these three are unique in that longevity was recorded. So what is the verdict? It is unclear. The Chinese and boys' choir cases revealed no difference in longevity compared to intact men; however, the Korean study recorded longer lives for eunuchs. Such is science.

In all fairness, even if these studies had been unanimous in their findings, they are not sufficient to allow for any solid conclusions. Many other factors are likely at play including the overall health and energetic status of the eunuchs, common population variation, and other factors that could affect longevity such as co-morbidity of other diseases. Given that the influence of testosterone elimination on aging and longevity is somewhat murky, we can now turn to the other end of the spectrum: testosterone supplementation.

Ornithologists in particular love to juice up their experimental subjects with testosterone. Some of my best friends are bird biologists, so I say that with total affection and respect. Ornithologists John Wingfield and Ellen Ketterson have had stellar careers looking at the effects of testosterone manipulation on the behavior and life history of various species of birds. Wingfield's particular focus species has been the Dark-eyed junco (*Junco hyemalis*).[13] I've spied them in my own backyard from time to time. They are handsome little birds. Wingfield and Ketterson have shown that increasing testosterone levels often improves a male's ability to establish and defend multiple nests, ward off competitors, and sire many more offspring compared to males that were not supplemented. Moreover, Wingfield noted that some males had naturally high testosterone levels while others did not. If testosterone was so advantageous for reproductive fitness, why didn't all males maintain high testosterone levels? Wingfield suggested that there is probably a cost associated with such high levels. He was right. In Juncos, testosterone-supplemented males had greater reproductive fitness but paid the price when it came to survivorship. Supplemented males put on less fat and had a harder time making it through the winter.

Wingfield therefore came up with the "Challenge Hypothesis," which suggested that there was a minimum level of testosterone necessary to maintain fertility and that any increase above this minimum was a reflection of increased reproductive effort when the reproductive benefits outweighed the survivorship costs.[14] This hypothesis has been tested numerous times in various bird species and even in primates including chimpanzees.[15] Overall, the hypothesis is supported. But what about humans?

Testosterone levels in men do increase in response to competition. This has been shown in numerous circumstances and in various cultures.[16]

Clearly testosterone is responsive to social and environmental conditions and often increases to levels that are far higher than what is necessary to maintain fertility. But why? The most common explanation is that testosterone promotes aggressive behavior and primes individuals to engage in competitive behavior. This is perhaps the case, but there are problems with this theory. Acute changes in testosterone do not increase the chances of aggressive behavior, but they can improve mood and increase confidence.

Our friend Bob Sapolsky and his collaborator Larry Tsai[17] offered another suggestion as to why testosterone levels increase beyond what is needed to maintain fertility. They wondered how muscle cells would respond to exposure to testosterone. To their surprise, they observed that muscle cell metabolic rate increased by 20 percent within two minutes of exposure to testosterone and stayed higher than controls for several hours. They published their results in a small journal devoted to the study of aggressive behavior. Not surprisingly, the journal is called *Aggressive Behavior*. Tsai and Sapolsky made the astute observation that competitive interactions in animals often involve posturing and preparation for a physical altercation, even though it may not come to fruition. Nonetheless, a male should be prepared to react swiftly and decisively, deploying muscle action in short order. Perhaps testosterone increases prime males for muscle action. Clearly acute and likely long-term increases in testosterone impose significant metabolic costs.

Do we have evidence of detrimental effects of testosterone in human males? Some is emerging. Many, if not most, men who take testosterone supplements do not exhibit any symptoms of hypogonadism. In one study, over half the men who were prescribed testosterone supplementation did not exhibit any symptoms consistent with hypogonadism or any other condition that would warrant testosterone supplementation.[18] It would seem in many cases that testosterone is prescribed in the same manner as vitamins, although the potential effects of anabolic steroids such as testosterone are much more powerful. Therefore, men who supplement are interesting subjects in a seminatural experiment. It is too soon to determine whether men who take testosterone supplements have shorter life spans, but some evidence is coming to light. Older men taking testosterone were much more likely to experience a serious myocardial infarction within a few weeks compared to men who did not start hormone treatment.[19] Earlier we posited that higher testosterone levels might be beneficial for muscle growth and development but that other organs might not be able to tolerate such levels. There seems to be a grain of truth here.

As a scientist, I am very curious to see how these testosterone-supplemented men fare over the long term. As a person, I wish them well. While there are those who champion the positive effects of testosterone supplementation on mood, confidence, and overall well-being, the experimental animal literature strongly indicates that there will be costs. Perhaps men with low immunological burdens and readily available food resources will be just fine. However, those who do not share these good fortunes or are older and physically compromised may need to think twice. Being healthy is more than feeling good. However, quality of life is something that I certainly value. There are no easy answers, but it is clear that male health involves weighing the costs and benefits of personal health decisions. I guess being healthy is more than hoping for happiness from a dab of gel from a doctor's prescription. It is being empowered by and making informed decisions.

SICK AND TIRED

When my tenure case was being considered at Yale University, I was fortunate to have Professor Stephen Stearns as an advocate. His summation of my research to me is captured by one phrase: "Macho makes you sick." Indeed. My own research and that of my former graduate student Michael Muehlenbein of the University of Texas at San Antonio have illustrated that human and non-human primate males bear a significant handicap by virtue of the Y chromosome. The Hollywood image of the swaggering, dashing man dispatching bad guys and carrying the day conjures up a perception of indestructibility. While men are on average larger and physically stronger than women, men have a considerable weakness. We have a harder time fighting off infections and illness compared to women, and as Dr. Porche rightly stated, men simply do not take care of themselves. This has a significant negative impact on the pace at which men age. Having a male physiology and all the bits and pieces associated with being male probably contribute to our shorter life spans compared to women. Nonetheless, being a non-smoking, non-drinking, non-speeding, non-coffee-drinking, hypochondriac homebody will not prevent insurance companies from charging you more for life insurance compared to an equally safe-playing woman. Male warranties run out sooner than those of women, often as the result of infection and other illnesses that affect men more than women.

It is ironic that machoism and the widespread impression that males are the tougher sex have persisted throughout history. Women have been seen as needing extra care and attention because of their fragile constitution, both mentally and physically. In contemporary times, it is not difficult to find a culture where females are perceived to be the weaker sex. In reality, there is stark evidence to suggest that if one were to use survivability as a benchmark of fragility, males would fall short of the female norm. From fetal development to old age, males exhibit a fragility that reflects a compromised capacity to deal with environmental stresses. Indeed, sex ratios reflect the fragile biological nature of human males. More male infants are born simply because more male than female babies are likely to die during infancy.

There are several potential underlying causes for these differences between males and females. First, there may be a collection of male-specific pathogens that tend to target men rather than women. There is very little evidence of this. Second, it may be that males are exposed to more opportunities for infection than women are. There is an air of truth to this in that some specific activities, such as hunting in hunter-gatherer societies, expose men to pathogens. Last, it may be that men are handicapped compared to women when it comes to fighting off infection. There is significant evidence that supports this idea.

Immunological function in men is often not as robust as it is in women. There are many reasons for this, some mechanical, some evolutionary. Testosterone suppresses immune function in important ways. In contrast, estradiol, a sex steroid that is also present in men but commonly much higher in women, bolsters immune function. Interestingly, and unfortunately, the effects of estradiol seem to be behind illnesses that are more common in women, such as autoimmune disorders. Diseases such as lupus and rheumatoid arthritis are autoimmune disorders, meaning the biochemical vanguards that usually defend against infection become too aggressive or lose their ability to distinguish between cellular friend and foe. A common treatment for these diseases is to lower estradiol levels in order to attenuate immunological function.

In wild populations, the negative effects of testosterone are evident in numerous species, including many birds, reptiles, and mammals. In most of the experiments conducted on wild animals, testosterone is increased or removed surgically or chemically while simultaneously manipulating some pathogen exposure. In the vast majority of these experiments, testosterone compromises immune function, stimulates more aggressive infections, and occasionally increases mortality. Does this occur in men? The evidence is sparse, but what is known supports this general response pattern.

A more rigorous test would be to remove testosterone from a man's system, infect him with a pathogen, and then reintroduce testosterone in varying doses. While slick and effective, it is unlikely, for good reason, that such a research protocol would get past university committees that protect the rights and well-being of people participating as research subjects in experiments. In such situations, scientists often turn to "natural experiments": situations in which people are subjected to a condition in everyday life that is somewhat similar to what would be created in a laboratory setting. In this case, men living in regions of the world where they would be likely to contract an infection during the normal course of their lives might be suitable research subjects. In one experiment that looked at the effects of malaria, an unfortunately all too common pathogen, on testosterone levels in Honduran men, researchers found that testosterone levels were lower in men with malarial infections but exhibited a rebound in response to malarial treatment.[20]

Since testosterone levels decline with age in some men, is there a concurrent increase in the ability to fend off infection? Not really. Much of what goes into immunological defenses involves energetics or the ability to afford and rally a response to an infection. As men get older, their ability to rally these defenses declines regardless of what happens to their testosterone levels. Western men who tend to have higher testosterone levels also have enough energy to maintain a robust immune system despite the potential detrimental effects of testosterone.

THE UNGUARDED SEX

When scientists first began to note that males tended to have shorter lives compared to females in a number of organisms, including humans, it was hypothesized that the difference in life span was grounded in genetics. In humans, males are the heterogametic sex. That is, their sex chromosomes are composed of two different chromosomes, X and Y. It was therefore suspected that, since the Y chromosome is relatively small compared to the X and does not carry complementary genes, detrimental genes on the X chromosome remain "unguarded" and more easily expressed. This is commonly observed in sex-linked disorders such as color blindness and hemophilia. Males are more likely than females to be afflicted with either of these disorders since the Y chromosome cannot mask the detrimental gene on the X chromosome. In females, the second X chromosome often serves to mask the detrimental gene on the complementary X. This theory has gained some traction through comparative studies. In birds, for example, females are the heterogametic sex ZW and express similar disorders resulting from the lack of a masking sex chromosome. However, as evolutionary biologists Alexei Maklakov and Virpi Lummaa note, mammals and birds tend to have very different social and mating systems that can confound this conclusion. They also note that the "unguarded chromosome" hypothesis has seldom been tested under controlled laboratory conditions, and when it has, the results have not supported the masking effects of the X chromosome.[21]

Another possibility has centered on mitochondria, the energy-producing organelles that keep our cells alive. Mitochondria are almost exclusively maternally inherited. It is therefore believed that mitochondrial genes that are beneficial or neutral to females but deleterious to males can accumulate and be detrimental to males' life span. In other words, because of maternal inheritance, males have no "say" in screening genes that may be harmful. If they are beneficial or neutral to females, they will remain, whether the males want them or not. This is a compelling idea; however, the evidence is a bit sketchy. Plus, while mitochondrial genes are immune to selection by males, nuclear genes that affect the expression of mitochondrial genes are not. Differences in accumulated damage due to oxidative stress within mitochondria are also a possibility in explaining the life span differences between males and females.

CANCER AND AGING: A CONNECTION WITH YOUNGER TIMES?

Few words evoke fear like "cancer." Medical advancements have addressed numerous health issues, but effective long-term treatments and preventative measures for cancer remain elusive and expensive. However, cancer is not one illness but a collection of many types of uncontrolled or malfunctioning cell growth. Among men and women, reproductive cancers are insidious in their own way since they are directly linked to our reproductive biology, which makes them all the more difficult to treat and prevent. Indeed, reproductive cancers are among the few illnesses that can be sex specific and tied to biological characteristics of being either male or female, although significant overlaps exist. In women, breast and ovarian cancers are primary concerns. The evolutionary biology and etiology of reproductive cancers that are specific to women are beyond the scope of this book, but there are several excellent discussions available about women's reproductive cancers.[22]

There is also a broad range of prevalence of various types of reproductive cancers in men, suggesting that environmental factors play a significant role in cancer risk. What can be inferred is that cancer was unlikely to have been a major source of mortality during human evolution since it is all but certain, in both men and women, that modern environmental conditions and lifestyle are major contributors to reproductive cancer risk. In men, breast cancer is relatively rare compared to women. However, the risk of developing breast cancer is greater for men who have Kleinfelter Syndrome (having an extra X chromosome, XXY) and greater exposure to estrogens. Moreover, male breast cancer is strongly correlated with female breast cancer at the age of fifty, whereas the risk of male breast cancer is not related to female breast cancer before that age. The implication is that factors associated with a lifetime exposure to carcinogenic agents contribute to both male and female breast cancers. There is a wide range of variation in the prevalence of male breast cancer around the globe.[23]

Testicular cancer and penile cancer are also somewhat rare compared to other forms of carcinoma. The prostate takes center stage when the topic of cancer arises in reference to men, especially as they age and enter the latter decades of life. So what is the prostate? In basic terms, it is a small gland, about the size of a walnut, that resides roughly in between the inner base of the penis and the anus. It surrounds the upper reaches of the urethra, sitting just in front of the urinary bladder. Its function is to produce fluid that supports sexual function, including seminal fluid. The evolutionary biology of the prostate is somewhat unclear, but some headway has been made. Recent genetic analysis of prostate cancer cells has suggested that the evolution of the genes that make men susceptible to prostate cancer evolved through a form of "punctuated evolution." That is, the genes associated with prostate cancer evolved rather suddenly as opposed to a gradual shift commonly assumed with regard to other traits.[24]

The incidence of prostate cancer is not evenly distributed across populations around the world. It is fairly well established that prostate cancer (as well as other cancers) is more prevalent in more developed parts of the world.[25] The reasons for this nonrandom distribution are still not clear. Various environmental and genetic factors may be at play. For example, a genetic variation that is strongly associated with prostate cancer is evident in some African and European populations.[26] Other factors that are sometimes mentioned include suspected environmental carcinogens such as sex hormones in water and food supplies. Sex steroids are commonly administered to dairy cattle to increase milk output and beef cattle to hasten muscle growth. While the steroid treatment of domesticated animals may be detrimental to the public health, it is through the increased risk of infection and the subsequent need to also administer antibiotics prophylactically that contribute to greater microbiotic virulence and resistance. Not much exogenous steroid actually makes it into the food we eat and into our systems.

A prime suspect that may explain the broad range of prostate cancer risk is testosterone and associated androgens such as five alpha dihydrotestosterone (DHT). Note that I did not say "testosterone levels" since there is virtually no evidence to suggest that an individual's testosterone levels can predict prostate cancer risk. However, the absence of evidence within individuals does not mean that when one pools a number of individuals, say, from a given population, that patterns do not emerge. It is difficult to ignore the fact that populations that have the highest testosterone levels also tend to have the greatest risk of and highest mortality from prostate cancer.[27]

Is there prostate cancer among non-Western populations? It is difficult to say since many parts of the world do not have the diagnostic capability of detecting prostate cancer or its related disorder, benign prostatic hyperplasia (BPH), which is a swelling of the prostate that sometimes, but not always, signals an imminent risk of prostate cancer. Anthropologist Benjamin Trumble and his colleagues found that in contrast to men from more Westernized societies, Tsimane men exhibit half the prevalence of BPH. They also had 68 percent smaller prostates compared to American men. Interestingly, even though Tsimane men exhibit much lower baseline testosterone levels than do American men, Tsimane men with higher testosterone levels were more likely to experience BPH.[28]

Similarly, anthropologist Ben Campbell at the University of Wisconsin, Milwaukee, made the intrepid attempt to quantify, if only in a coarse manner, the incidence of prostate health problems among Ariaal cattle herders of northern Kenya. Using interview survey data on general health, Campbell found that complaints consistent with prostate health problems rose significantly in men over the age of sixty. Although he surveyed Ariaal men who were sedentary compared to those who were nomadic, there were no differences in the rates of prostate hyperplasia symptom complaints between the two groups.[29] As Trumble points out, BPH is probably not an inevitable outcome of aging. Lifetime exposure to higher endogenous testosterone levels in association with greater energy intake in Westernized populations probably plays a major role in the emergence and prevalence of prostate disease in the United States and elsewhere. Does this mean that high testosterone levels lead to prostate disease or cancer? The answer is not simple and is nuanced. There is very little evidence that testosterone predicts prostate cancer risk in individuals living in Westernized societies. However, differences in testosterone exposure between populations may provide evidence of a more complex association with prostate disease.

The relationship between testosterone and prostate cancer risk was cleverly elucidated by Louis Calistro Alvarado from the University of New Mexico. Instead of simply correlating testosterone levels with risk, he compared proportional disparities in testosterone levels with prostate cancer risk disparities between different ethnicities within specific communities. In this way, he was able to control for between-population variation in energetic availability and assess actual risk differences between communities that exhibit various ranges of risk.[30] His findings revealed a strong correlation between the range of disparities in testosterone levels and the range of disparities in prostate cancer. Other investigations have shown an increase in prostate cancer in succeeding generations of immigrants, especially from Asia. While better cancer screening may underlie the detection of more incidences of cancer, there is also the likelihood that changes in lifestyle, including diet, may be causal.[31]

Why has natural selection not provided men with the ability to maintain good health throughout their entire lives?[32] If women can have robust immune systems, why can't men? The answer lies in life history trade-offs and constraints on the biology of men. There are some obvious ways in which men would be buttressed against infection, such as better nutrition, but there would still be a disparity between men and women. We can try to approach this by using a mental experiment.

We know, for example, that estradiol can bolster immune function, and it is quite easy to create estradiol from testosterone, a hormone that is in ample supply in most men. What if natural selection had allowed men to have higher levels of estradiol? This would almost certainly have had a positive effect on a man's ability to fend off infection. The downside is that there would be undesirable effects, not the least of which would be compromised reproduction. Estradiol would shut down the hypothalamus, fertility, libido, and the ability to build and maintain sexually dimorphic muscle mass. This is not something any male would want. Or is it? As men age, estradiol levels do tend to increase, mostly as a result of greater aromatization and fat mass. Since men can expect to live well past their fifties, there should be some strategy that men can deploy to deal with the constraint of losing testosterone and gaining estradiol. Perhaps unavoidable increases in estradiol due to age-associated changes in reproductive endocrinology were leveraged to support male longevity and other reproductive strategies such as paternal and mate investment. Getting older and being more susceptible to infection and illness are probably inevitable; however, there seem to be strategies by which men can buy a bit more time to increase their fitness.

HOMO "ERECTUS"

Most cannot say the words without giggling, blushing, or otherwise feeling like an immature twelve-year-old. But that's okay. It can be done but only after years of serious postgraduate training and a healthy respect for paleoanthropology. Say it. Homo erectus. Yes, it literally means "erect man." Homo erectus was a hominid ancestor who contributed to the evolution of modern humans, otherwise known as our own species, Homo sapiens. The last two million years or so was an important period during our evolution, one that has spawned decades of important research and numerous books. Indeed, one of my esteemed Yale anthropology colleagues and his collaborators discovered what could be the earliest fossil evidence of our genus Homo.[33] We will indeed be talking about the common sophomoric interpretation of these two words. The topic of discussion will be men and their erections, with special emphasis on how erections (or lack thereof) are part of the conversation surrounding aging men.

Without the aid of fertility specialists, petri dishes, and expensive medical instruments, most conceptions occur the old-fashioned way. On the male side of things, the penis acts as an instrument of stimulation and pleasure, optimally for both parties. However, mechanically, the penis is a delivery device that has evolved to deposit gametes near the site of fertilization, usually at or near the cervix so that sperm can make their way to the uterus and, ultimately, the ovum. To accomplish this task, the penis needs to change from its common flaccid state to one of rigidity. This is especially important since humans are the only primates that do not have a penis bone. How and why we lost, or ever had, this anatomical feature is another conversation. Mostly because I don't have a clue. Querying colleagues who are experts in primate evolution about this topic results in a collective hapless shrug. Perhaps I will have an answer in my next book. Nonetheless, this change allows for more efficient deposition of sperm and a pleasant time for all. With age, however, this process can unfortunately be compromised.

Erectile dysfunction (ED), in the past called impotence, is when an erection is not possible or difficult to achieve. Mechanically, for an erection to occur, very small muscles surrounding blood vessels near the base of the penis change their state. Some relax and allow blood flow into the penis; others contract and restrict the flow of blood out. The result is a fluid logjam that traps blood in areas of the penis, primarily the corpus cavernosum, resulting in increased blood pressure and tissue rigidity. From a neurological perspective, an erection can be stimulated by virtually any sensory stimulation, although the usual suspects are visual, auditory, and tactile. Signals from the brain travel down the spinal cord and link with nerves in the base of the spine that then signal the appropriate blood vessels at the base of the penis. Following orgasm, these blood vessels return to their state of relaxation, blood is allowed to circulate more freely, and the penis returns to its flaccid state. Drugs such as Viagra act on the blood vessels by altering levels of nitrous oxide, the chemical that is responsible for changing the state of the small muscles surrounding the vasculature that control the onset of erections.

With age, this process tends to become compromised, although the reasons are many and need to be sorted out. To start, there is what is called the "honeymoon effect." Couples tend to have more sex early in their relationship. With the passage of time, checking the cable lineup or finishing that mystery novel tends to take on greater importance resulting in less frequent sex. The reasons for the decline in sex frequency are probably related to boredom, changes in attractiveness or perceptions of attractiveness, life stress, fatigue, or any number of life distractions that can get in the way of having a good time. Setting those issues aside, the chances of a male having difficulty inducing and maintaining an erection increase with age, especially after the age of forty. There are many physiological reasons for this including aspects of aging that compromise cardiovascular performance and neurological function. These aspects in tandem can obviously have adverse effects on sexual function.

Changes in hormone levels, specifically declines in testosterone, can reduce libido, but there is considerable variation in this association. For example, one study has shown that erections and sexual function were maintained with modest testosterone levels by modulating these levels within the clinical normal range.[34] Men with lower testosterone levels do not necessarily have compromised libido. When using hormone measurements to determine a man's ability to carry out any libidinous duty, proceed with caution. In the same way testosterone levels vary across populations, we should be aware of variation in ED across populations and, indeed, perhaps in other species.

Ellison and colleagues' study on differences in testosterone changes with age illustrates the importance of being aware of the broad range of variation that is often the norm in human evolutionary biology.[35] It would be interesting to discover whether similar problems with ED are common in populations other than Western ones or if the prevalence of ED is simply something marketed by pharmaceutical companies to drive sales among American men attempting to recapture their youth. The answer seems to be that age-related patterns of ED or symptoms that are likely to be indicative of ED are pretty common across populations and cultures. This should not be much of a surprise since homemade and native male potency remedies are common across many different societies.

In South America, *Lepidium meyenii walp* and *Jatropha macrantha* are root vegetables that are indigenous to the Peruvian Andes and purported to increase male vigor, loosely defined. Scientific evidence suggests that these roots induce significant increases in serum testosterone in male mice. Both also stimulate increases in estradiol and progesterone in female mice, indicating that they act as pseudo-gonadotropins, or brain hormones, that stimulate steroid hormone production in both the testes and ovaries.[36] Regrettably, animal products derived from tigers, rhinos, and other endangered animals not only fuel the tragic decline of these rare and magnificent species but also speak to the often misguided worldwide male market for medicines and cures for erection problems. To those men who will forget everything else in this book, mark this page. You may want to look into massive amounts of garlic and casein, a very common protein derived from dairy products and found in your local health food store. It stimulates a very significant increase in testosterone levels in male mice.[37] Leave the rhinos and tigers alone.

ED is common across populations in the Americas, Europe, Asia, and Africa, although there is a broad range of variation in its prevalence, especially at older ages.[38] Evidence for ED is even evident among remote cattle herders in Africa. Anthropologists Peter Gray and Ben Campbell reported that among Ariaal men of northern Kenya, men over sixty exhibited significantly higher rates of ED compared to their younger counterparts. Interestingly, after controlling for age, the risk of ED is negatively associated with the number of wives a man has. In other words, the more wives a man has, the less likely he is to exhibit ED. At first glance, this may be an example of the Coolidge Effect, a phenomenon in which vertebrate males are able to recover more quickly from bouts of mating when there are multiple different females involved compared to males exposed to the same female multiple times. However, before the male readers with ED or those wishing to ward it off go spouting the benefits of extramarital affairs and chasing anything that moves, there are important caveats. Among the Ariaal, men with multiple wives have more wealth in the form of cattle and in general tend to be healthier and more physically robust compared to men with one wife or no wives. Wealth, health, and number of wives tend to go hand in hand among the Ariaal.[39]

How common is ED in developing populations? Is it simply a condition of the modern world? There are a few studies of ED from Europe, Asia, and Latin America. All basically report the same results. That is, men over the age of forty tend to exhibit various degrees of ED quite regularly.[40] In less remote parts of the world, ED is not unusual and consistently associated with aging. A meta-analysis by Prins and colleagues[41] demonstrated that the risk of ED tended to increase with age regardless of population. However, the incidence of ED exhibited a broad range of ED risk at older ages. After sorting through studies that examined the incidence of ED by age and controlling for other potentially confounding factors, these researchers found that the range of variation in ED risk in men younger than forty was low. In men older than seventy, the range of risk was quite a bit higher, and higher still in men older than eighty. The authors attempted to control for methods of assessment and other extraneous factors that may add noise to the data. However, it is fairly certain that the broad range of variation at older ages is surely related to lifestyle, mating patterns, and other biocultural factors.

For all intents and purposes, there are no data on and there is no evidence of erectile dysfunction in other primates. One study of chimpanzees has shown that males exhibit a fair degree of variation in erectile scores in response to exposure to a sexually mature female during various phases of her menstrual cycle. Although the sample sizes in this study were small (five males, one female), they were somewhat informative. While it was clear that males were more attracted to and sexually aroused by females that were ovulating and exhibiting sexual swellings, male erection scores varied significantly, suggesting that different males exhibited different abilities to generate an erection and/or their level of sexual arousal varied for reasons that remain unclear.[42]

SILVERBACKS, BALDING, AND CESAR ROMERO

Gaining weight and losing muscle may not signal the end of an older man's ability to reproduce and attain additional fitness. However, other changes occur that are energetically less costly but have the potential to affect an older male's attractiveness to potential mates as well as signal his age. Compared to other primates and mammals, humans have very little hair, but what we do have tends to serve as a signal of age, reproductive status, and an adaptation to cope with the environment.[43] Many of the traits that accompany aging are common to both men and women, most noticeably the loss and/or graying of hair.

I'm one of the fortunate ones. Thanks to my parents and other ancestors, there is little chance that I will experience male-pattern baldness. I admit that there has been a bit of thinning, but I'm fairly certain that I will have a full head of hair when this life has run its course. However, the whole gray thing is another matter. I now have more gray than black, but that suits me just fine. As one of my professor friends reminds me, it gives us male and female faculty a bit of gravitas. But my acceptance of gray hair is not universal. Television and radio commercials pander to aging males, urging them to do something about their receding hairline (as well as their testosterone). Male aging is often typified by alopecia (hair loss), changes in hair color, and the growth of hair in some inconvenient places. Is this unique to humans? This is difficult to say since humans are the only primate that is relatively hairless. Nonetheless, if you are reading this and you are an older male, you know exactly what I am talking about. If you are socially conscious, you probably own a pair of tweezers and have a favorite pair of scissors to trim those unseemly hairy bits.

Changes in hair color, however, are somewhat more ubiquitous across mammals. If you have ever owned a dog or cat with a dark muzzle that lived to a ripe old age, you will remember the whitening of the fur around the nose and eyes. Hair color change involves the diminishment of melanin that gives hair its dark color, hence the whitening of head and beard hair. The cause of melanin loss was a bit of a mystery until recently. We can be fairly certain now that gray hair is caused by oxidative stress, the same mechanism that was the topic of discussion in chapter 3. In essence, the buildup of hydrogen peroxide resulting from the breakdown of oxidative stress protection mechanisms causes the degradation of melanin.[44]

So how does this affect a male's reproductive success? This is unclear, although if females were to preferentially be attracted to older men, gray hair would be a fairly accurate signal that a man is in their desired age range. Why would a female be attracted to an older male? It might be that an older male has demonstrated the ability to outlive many of his competitors, perhaps indicating good genes that a female may wish to be passed on to her offspring. Interestingly, male chimpanzees tend to prefer mating with older females who have already reproduced, perhaps as a way of ensuring that they are investing mating effort in a female who has proven fertility compared to a younger female who has not reproduced yet.[45] Extrapolating this idea to older males is not then that far-fetched, especially since physical vigor may take a backseat to other traits when it comes to the evolution of the mating game during human evolution.

Facial hair, the growth of which is stimulated by testosterone, is sometimes associated with age, although clearly younger men can and do grow beards and mustaches. In a cross-cultural study involving women of European descent in New Zealand and Polynesians from Samoa, subjects viewed images of bearded and clean-shaven men and were asked to rate their attractiveness. Neither population found the bearded men to be more attractive, although clean-shaven men were judged to be moderately more attractive. However, when women viewed images of men exhibiting an aggressive facial expression, the ones with beards were judged to be more aggressive, to be older, and to have higher social status compared to clean-shaven men. The authors concluded that facial hair may be useful for augmenting aggressive stances toward other males, which may have a positive influence on fitness independent of attractiveness.[46]

Hair loss, or alopecia, is another matter. There are many causes and forms of alopecia, many of which are present in both men and women. However, alopecia associated with aging is more common in men, although some of the same mechanisms that are causal in men are present in women. The cause of alopecia involves many factors including, again, hormones. In women, alopecia is commonly associated with obesity and diabetes. When a woman becomes insulin resistant, high levels of insulin stimulate the ovaries to produce lots of testosterone for conversion to estradiol. However, with hyper-insulinemia, so much testosterone is produced that it outpaces the ovaries' ability to convert it all to estradiol. The excess testosterone then makes its way into the circulation and contributes to alopecia and hirsutism (facial and other body hair growth).

Similarly, men are exposed to androgens throughout their lives. The two primary hormones involved in hair loss are testosterone and DHT. DHT is a sort of super-testosterone formed through the conversion of testosterone by the enzyme five alpha reductase. Most DHT production and receptor binding are localized in the prostate so very little makes its way into general circulation. However, some men have more receptors for DHT in their scalp. As men age, the cumulative exposure of hair follicles to DHT contributes to hair loss. Some hair loss remedies such as finasteride are DHT blockers meant to reduce hair loss associated with androgenetic alopecia. This is the reason women are discouraged from handling many hair loss medications; these substances can cause serious problems in male sex organ development in fetuses.[47]

One would be hard-pressed to develop an adaptive explanation for male alopecia, especially since the mechanisms are now well understood and shared by women who experience the same hormonal changes. Nonetheless, alopecia could potentially serve as a signal of a man's age and perhaps his hormone status, although we would be standing on shaky theoretical ground. Since other primates and mammals experience hair loss with age,[48] it is likely that alopecia is a simple constraint common to most if not all mammals.

GAY AND GRAY

Sexual orientation and identity are crucial aspects of health and well-being. Yet here is a glaring gap in this book. There has been no mention of gay men. There are no bisexual men. There are no transsexual men. I will try to rectify those omissions here, but it will be challenging since this area of biological anthropological research is pretty undeveloped. Although there are numerous studies that examine the sociocultural aspects of gay older men,[49] evolutionary perspectives have been proposed with much less frequency and more than a little bit of justifiable caution.[50] We probably know more about the evolutionary biology of male aging in our closest cousin, the chimpanzee, than we do about this important segment of the human experience. As I mentioned previously, the scientific history of biological anthropology is checkered, with many sexist, classist, and racist conclusions wrapped in veils of quasi-scientific research. I think biological anthropologists are hesitant to repeat any mistakes that would breathe new life into these discarded ideas, even if it were done inadvertently.

Sexuality, in all its forms, is the result of natural selection, or at the very least influenced by other aspects of human life history biology that have resulted from evolution by natural selection. For example, straight or gay, the neuroendocrine processes that give you goosebumps and an erection at the site of an attractive person are the result of natural selection. The hormones that facilitate pair bonding between a man and a woman, or two men, are the result of natural selection.[51] Most studies assume that aspects of life history are anchored in heterosexuality, but it is becoming increasingly obvious that alternate sexualities have been around a long time and are important in understanding every aspect of the evolution of human life histories, including how men age.

In their effort to understand the hows and whys of sexuality, human evolutionary biologists often view heterosexual behavior as having evolved in response to the need to develop sexual strategies, that is, optimal behaviors that result in greater reproductive success, in both men and women. Similarly, it would be constructive, and downright interesting, to put gay men under the same lens and investigate how this aspect of human sexuality evolved and how it changes throughout the human life course. How does the life of a gay man change from the time when he is full of health and vigor in his twenties to when his hair starts to thin and gray after the age of fifty? How does population variation play into understanding the aging process in a gay man? Indeed, is homosexuality even accurate to describe some aspects of gender identification and sexual preference? This topic merits much more detailed discussion than this chapter is capable of providing, but we can draw on some anthropological examples of variation in human male sexuality.

Among the Ache of Paraguay, most men self identify as wanting to engage in typical male behaviors, such as hunting and trying to procure a wife. However, there are a few men who have opted out of this life, instead choosing to spend most of their time with women and engaging in more typical female tasks such as cooking and child care. Among the Ache, there is a somewhat stark delineation between the tasks performed by men and women. While the men who lead a less "typical" life are accepted as part of the community, they are often not treated as well as others with more common male-gendered roles. In a similar way, marginalization of gay communities tends to occur much too often and likely has an effect on the health and well-being of this community as they age.

To those who argue that homosexuality is antithetical to evolutionary biology, remember that most sex is nonreproductive. In essence, sex often functions to facilitate bonding. This leads to interesting questions, such as, what are the hormonal mechanisms that facilitate this bonding in gay men? Do they change with age? Do sexual preferences in partners change with age? Are older gay men more monogamous?

From a comparative perspective, we know that other great apes engage in homosexual behavior on a regular basis. Bonobos, also known as pygmy chimpanzees, are well-known for their wide range of sexual behaviors as well as their use of sexual behavior as a form of social bonding and political negotiation and as a means of addressing social tensions. Among bonobos, females are most well-known for their genito-genital (GG) rubbing, although males also engage in homosexual behavior. Under certain conditions, common chimpanzees also engage in homosexual behavior, suggesting that sexuality is malleable in relation to ecological or social contexts.[52] However, does this change with age? This is largely unknown mostly because it is notoriously difficult to know the ages of great apes in the wild. Do other great apes such as orangutans or gorillas engage in homosexual behavior? To my knowledge, this has not been observed in the wild, although such information from wild populations is extremely difficult to obtain. What we do know is that humans operate on a sexuality spectrum that goes beyond simple gay/straight dichotomies. Understanding this from an evolutionary and life history perspective would be useful for gaining health insights.

We might start by asking older gay men how they view the aging process. In one of the few studies that address this issue, Robert Schope of the University of Iowa surveyed both gay men and lesbians and found that gay men tended to have a more negative view of aging compared to gay women. The reason for this is not clear, but men's responses tended to focus on a fear of being less attractive and on the fact that social workers should be aware of the needs of this segment of the community.[53] Although this is just one study and does not necessarily reflect the perceptions of all older gay men, it does serve to bring greater awareness of the need for more gerontological research that would address the health needs of older gay men.

As discussed earlier, sex and sexual behavior likely affect mortality and morbidity in men. Early demographic studies of homosexual men during the AIDS epidemic suggested that male homosexual life span was shorter and mortality was significantly higher than those of heterosexuals. Given that this was at the height of the AIDS health crisis, these results are probably not surprising.[54] But as is the case for heterosexual men, risks associated with sexual behavior are likely to contribute to variations in morbidity and mortality for homosexual men. Some evidence suggests that these risks are highest during young adulthood, which is when we see the bump in mortality. Since demographic studies seldom collect data on sexual orientation, it would be reasonable to assume that a significant proportion of the men in these studies are not heterosexual. They are therefore probably contributing to this demographic mortality "bump" during early adulthood. How is this manifested? The portion of the mortality bump attributable to homosexual men is probably due to habits that are similar to those of heterosexual men, particularly high-risk behavior (e.g., smoking and drinking) and other unhealthy activities. Again, these factors are the same as those for heterosexual men so there is no reason to assume that homosexual lifestyles are somehow inherently more risky than those of heterosexuals. This does not mean that sexual orientation cannot potentially introduce specific health risks as was tragically evident during the AIDS epidemic when the gay community suffered disproportionately compared to heterosexual populations. As a whole, sex can be dangerous no matter who you are.

A recent Danish study looked at mortality patterns in lesbian and gay same-sex marriages to gain a greater understanding of previous demographic assessments that suggested that gay men and women tended to have higher adult mortality compared to the majority heterosexual population.[55] The study consisted of over 8,000 individuals, of whom 4,914 were men. Male mortality in the study group was 165 percent higher than that of men from the general population after 1–3 years of marriage but declined to levels that were not significantly different from those of the general population after 13–14 years after marriage. The mortality difference immediately following marriage is not fully understood although the researchers suggest that preexisting conditions such as HIV infection prior to the subsequent introduction of highly active antiretroviral treatment (HAART) may be contributory.[56] After 1995 and the introduction of HAART treatment, mortality was about 33 percent higher in married homosexual male couples compared to age-controlled married heterosexual men. Similar patterns of mortality were noted in lesbian couples, which suggests a more robust influence on the same-sex marriage community. So what does this tell us about the role of sexual orientation in male mortality and aging? Not much except that men in same-sex marriages may be at risk of higher mortality and morbidity. Since the gay community has been and continues to be a target of discrimination and social marginalization, it should not be surprising that this population may be at greater risk of stress-related health challenges as well as unhealthy stress coping behaviors such as drug use, alcoholism, and smoking.[57]

Although mental and physical health information on older gay men is sparse, a recent investigation reported that depressive symptoms in older and middle-aged gay men were more prevalent than in heterosexual men.[58] The most likely cause, the authors of the study argue, is that older gay men endure lifelong stress as a sexual minority and that this exacerbates the burden that is commonly the result of aging.[59] However, age-related stress appears to be alleviated in men in same-sex marriages, most likely as a result of a sense of empowerment and presence of emotional support from husbands.[60] Being accepted and incorporated into the fabric of a community seems to alleviate the stress of being a sexual minority.

In the end, the health of older men is multifaceted, often engaging trade-offs and constraints that are part of their evolutionary biology. Survivorship and investment in keeping body and soul together have evolved to be in service of reproductive effort. This is also true for women. Setting aside traumatic injury and many stochastic genetic disorders, to understand health, one needs to embrace the central place of reproductive function. From cancer to diabetes to many other challenges, one is hard-pressed to identify any illness, infectious or non-infectious, that does not engage reproductive biology in some way. In addition, appreciating the importance of human diversity in all its forms is vital for a deep comprehension of disease and illness. Until health professionals embrace this basic premise of evolutionary biology, the etiology and sources of many health challenges will continue to be elusive.

Despite our advances in broadening our understanding of male aging, much remains to be explored. We've only begun to touch upon the full range of human sexual identities and gender roles that emerged during the evolution of our species and distinguish us from other primates and mammals. Similarly, our understanding of human diversity and male aging has reached an adolescence that remains to fully mature. As we move forward with these and other questions, how we continue to evolve as a species looms as a much larger question. How this will continue to play out is uncertain although the contribution of older men to this future certainly spawns deep, enticing possibilities.

CHAPTER 7

OLDER MEN AND THE FUTURE OF HUMAN EVOLUTION

Old men are dangerous: it doesn't matter to them what is going to happen to the world.

—George Bernard Shaw, *Heartbreak House*

A 2014 Gallup poll reported that 42 percent of Americans believed that evolution had no part in the emergence of humans.[1] This is an astonishing and dreadful figure. Not only is evolution by natural selection central to understanding our past, but it will surely affect our future. No organism is immune to the effects of natural selection. What does change are the agents of selection. Death, morbidity, and mortality are primary agents of change. We will all be cleared out, perhaps as a species, surely as individuals. The question is how that change will unfold and how older men will be enmeshed.

In this concluding chapter I will suggest a final point that I hope will put a bow on what has been discussed so far. Not only is evolution by natural selection still relevant, but older men are likely to play a major role in shaping our evolutionary future. The salient question, however, is: What will that role be and how will our future be shaped? Before we delve into the particulars of older men, we will make the informed assumption that natural selection will continue to be relevant to evolution. Looking at the history of other organisms, we can draw some conservative predictions about the future of Homo sapiens. There is an oft-cited figure that

99 percent of all species that ever existed have gone extinct. That does not bode well for us. Sometimes species do not go extinct but evolve into something else. This has been a hallmark of human evolution since it is likely we evolved from earlier hominids. But assuming we do not go extinct, either by our own hands or a rogue asteroid, we can examine the emerging selective factors that will drive human evolution and how older men will be involved.

MODERN EVOLUTION

When people learn that I work with forager populations in South America, they sometimes ask if there are still uncontacted groups of people in the world. The answer is "sort of." It is difficult to imagine any population who hasn't seen an airplane or those odd moving stars that we know as satellites zipping across the night sky as they continue their equatorial orbits.[2] In fact, imagery from a few of those satellites reveals that there are still remote and uncontacted communities in the Amazon.[3]

This tends to amaze people who are then quick to ask if expeditions are being mounted to reach these communities or if there is some mandate to maintain the pristine nature of their existence. The fact is that researchers are hesitant to go out and make contact with these communities since there are many risks, not the least of which is exposing them to novel infectious agents that can quickly decimate a population that has little or no immunological defense against those pathogens.[4] Plus if those remote communities simply desire to be left alone, we should honor their wishes. Monitoring these communities is, however, good practice so as to lower the risk of unplanned and potentially disastrous contact with corporate-sponsored explorers seeking natural resources such as oil and timber, military incursions, or poor farmers simply looking for a place to cultivate their crops and call home.

Indigenous groups that are transitioning to modern economies and engaging with the predominant societies can and do benefit from the technology and tools that flow from market economies. For example, some older hunter-gatherer men now have access to eyeglasses.[5] This is still rare since eyeglasses are pricey and access to them is limited. But I have seen an increasing number of older men sporting prescription lenses. Where do they get them? Missionaries sometimes donate them. Other times older men get them from visits to local towns. This might seem like a curious but trivial observation, but it is one that is telling of the dynamic nature of natural selection and the importance of human innovation and culture.

In the not so recent past, developing bad eyesight as a result of aging was a serious setback, as it affects one's ability to hunt, forage, work, and generally cope with the daily needs of life. Those with terrible eyesight, like me, would have been at a distinct fitness disadvantage compared to those who could see clearly. Has technology finally trumped natural selection? Probably not, but it certainly will factor into the future of our species. In fact the emergence of basic tools to hunt and process food probably contributed to the physical changes that evolved in our species compared to earlier hominids. The benefits of cracking open a nut with a rock as opposed to running the risk of breaking a precious tooth are enormous and significant over evolutionary time. Technology and new tools will continue to play an important role in our evolution.

However, technology and modern innovations will not stop natural selection but will simply change the factors that drive nonrandom changes in fertility and mortality across populations. Recall, for example, how the effects of technology increased young male mortality and decreased the age of puberty.[6] These nonrandom changes in mortality have potentially significant selective effects on a population. This opens up the possibility of evolution. Technological innovations do not replace the natural environment; they add to it. As such, technological advancements simply become another agent of change.

Improvements in medical technology and the emergence of options to plan one's reproductive life, especially for women, have also had a significant impact on populations that could lead to evolutionary change. For example, declines in age-related mortality and the growth of older segments of the population that are outpacing fertility in countries like Japan, Portugal, and Greece put a strain on social and economic resources to care for these older individuals. Similar issues face the United States as the large cohort of aging individuals known as the "baby boomers," those born between 1946 and 1964, continues to retire and draw upon the nation's health care, economic, and social resources. Clearly populations that have declining fertility and a growing population of aging individuals may not be sustainable. How these demographic issues will play out remains to be seen.

The development of family planning technology also holds the potential to alter natural selection in human populations. Perhaps if women were able to guide their reproductive lives, humans would be the first species to take control of the reins of evolution. However, there are two constraints that inhibit this possibility. First, the availability of contraception technology is not widespread across human societies. While it is widely available in the United States and other developed nations, women in developing nations who arguably need it the most do not have regular and affordable access to it.[7] Second, older men largely control the availability of family planning care. This is especially true in developing nations. Older men literally control the reproductive lives of most women around the world. This empowers older men to have significant influence on human reproduction in many parts of the world.

Since older men tend to hold the reins of economic and political power in most societies, their role in controlling resources, including emerging technologies, and shaping human fertility and mortality in a nonrandom fashion is all but certain.[8] In other words, whatever evolution has in store for us, older men almost surely will have a major influence. In addition to having significant control over technology that directly affects human fertility, older men also control major influences on our mortality. Nowhere is this more evident than in conflict and warfare.

OLDER MEN ARE DANGEROUS

I have no military experience, but I have always had an interest in how the military is organized and how individuals interact, cooperate, and persevere in the face of extreme adversity and confusion. Organized conflict or warfare is a defining characteristic of human males. The only other great apes that engage in anything like war are chimpanzees.[9] In many ways, war and conflict are an important part of the story of human male evolution. What is interesting and tragic is that warfare has become more intense and lethal compared to what our pre–Homo sapiens ancestors experienced, largely because of the incorporation of technology and mechanization. Some have argued that conflict in many societies has decreased in recent history, but this does not address the intensity of warfare or the terrible potential of high-tech conflict such as the growing capacity for nuclear war.[10] Being big and strong can be an asset during war; however, it is no longer necessary. In many ways, warfare has become the domain of older men deploying the services and lives of younger men.

Anthropologists have a long history of studying warfare and conflict. As one might imagine, it can be dangerous work since it often leads researchers into volatile and remote areas, often asking questions that can be uncomfortable and risky for both the participant and the researcher. During the course of my career, I seem to have been drawn to these populations. For example, in the recent past, the Ache engaged in organized club fights with rival bands. Although I never witnessed these fights, I vividly recall the scars and dents in the scalps of men, evidence of healed skull fractures. Anthropologists Kim Hill and Magdalena Hurtado noted in their work on the Ache how the aftermath of these club fights resulted in miserable times for the community since women and children also suffered, not from direct attacks but from hunger since the men were often too wounded and weak to hunt for many days after a fight.[11] Although older men were often not participants in the fights, they were often involved in the political discussions that led to conflict. Older men can wage war by proxy through the actions and coaxing of younger men.

In my present research among the Shuar of southeastern Ecuador, there are many reminders of the role of war and conflict within this society. This region of Amazonia is somewhat notorious for long-standing conflicts, revenge killings, and important rituals associated with warfare that also include other tribes. Among the Shuar, the most prominent ritual is the *tsantsa*, or the taking and shrinking of an enemy's head. I do not profess to completely understand the significance or importance of this ritual, but I do know that it is not simply the act of taking a trophy.[12] The Shuar and Achuar people of Ecuador and Peru consider this practice to be more important and significant than merely recording the vanquishing of an enemy. Warfare and organized conflict have been woven into the cultural existence of these societies. Today many younger Shuar men serve in the Ecuadorian military, often in units that specialize in jungle warfare. Many Shuar and other indigenous men also participated in the Cenepa conflict with Peru in 1995.[13] This connection with warfare plays a central role in the life of many indigenous peoples. Acceptance and engagement in conflict does not mean that societies enjoy or relish war, but warfare and conflict are often seen as important, albeit painful, realities of life especially for men.

The role of older men in conflict has also shifted during human evolution. In the opening monologue of the film *Act of Valor*, one of the main U.S. Navy SEAL characters writes a letter to the son of his commanding officer, who has just been killed in action. He quotes his own father, writing, "The worst thing about growing old is that other men stop seeing you as being dangerous." Given my lack of military experience, I certainly cannot attest to understanding the sentiments of that comment from their perspective. However, now that I am over fifty years old and recognize that I'm not the same young man who made his way through the sometimes dicey neighborhoods of south-central Los Angeles during his formative adolescent years, that statement resonates with me just a bit. Other males viewing you as being at least a little bit dangerous can have its benefits. It creates a form of détente that allows you to get through the day without overt conflict. Often it is not so much appearing dangerous as it is not displaying fear. Older men are not commonly viewed as being intimidating or as being a threat. They are also less capable of engaging in the risky behavior and skirmishes that are part of a younger man's life. But older men have developed social and political strategies that make them dangerous in different ways and more efficient at incurring subtle and outright mayhem on a grand scale than were our evolutionary ancestors.

If the behavioral traits and decision-making processes associated with warfare have been subject to natural selection, then there should be evidence for nonrandom variation in reproductive success, associated traits that are heritable, and variation in those traits particularly among older men. One can argue that the mental capacity and neuroendocrinological potential for violence may engage the last two requirements for natural selection.[14] That is, the hormonal milieu of older men, the central role of competition in male life histories, and variation in risky behavior across the life course have likely shaped the neuroendocrine organization of men to make it more conducive to engage in conflict, including war. In no way am I insinuating that warfare and conflict are the inevitable outcomes of male evolutionary biology. The plasticity of our behavioral repertoires allows us to seek solutions that steer us away from conflict and war. However, this is an uphill battle, so to speak, in which a concerted effort needs to be made to avoid conflict. That said, we are still left with the issue of a key necessity for natural selection to take place: nonrandom variation in reproductive success. There are no clear answers, and only a handful of studies have explored this question.

The Nyangatom people are nomadic pastoralists in Ethiopia who regularly raid neighboring communities for livestock and to avenge a previous attack by rival villages. These raids, which follow a regular pattern of incursion and retaliation, often include the use of automatic weapons and can be quite lethal. Anthropologists Luke Glowacki and Richard Wrangham of Harvard University conducted a long-term investigation of this group and found that elders who were identified as prolific raiders when they were younger had more wives and offspring than other older men. Raiders were not more likely to come from families with fewer older maternal sisters or a greater number of older maternal brothers. Glowacki and Wrangham suggest that raiding may provide opportunities for increased reproductive success.[15] If the results of their investigation prove to be valid, it would suggest that engaging in warfare as a young man and surviving to old age have potential fitness benefits. As we discussed earlier, this aligns with the mortality bump observed in many young male populations and is consistent with the prevalence of high-risk behavior at younger ages. If a male survives this early gauntlet of risky behavior, there may be social and fitness benefits later on.

However, the suggestion that male involvement with organized warfare results in greater male fitness has some shortcomings. Data from other populations who engage in similar types of organized warfare such as the Turkana of Kenya suggest that other mitigating factors may be at play such as the underestimation of the mortality costs and the focus on "stealth" raids as opposed to more lethal "battle" raids[16] in the Nyangatom study, the difference being that battle raids result in more lethal exposure.[17] Stealth raids involve fewer men and involve less risk. Clearly while further research is needed, the prevalence of organized warfare in human societies is difficult to explain without a biocultural perspective that at least acknowledges the possible engagement of evolutionary processes.

Research among cattle herders like the Turkana and Nyangatom only get us so far in understanding organized conflict and the role of older males. These conflicts are mostly contained within a society's immediate geographic region and have little or no impact on people outside the community. Among industrialized nations, warfare is different. It is mechanized, semi-automated, and largely controlled and managed by older men who are nowhere near the battlefield and have the potential to influence a much larger number of people, most of whom are noncombatants. Modern warfare is still largely the domain of men and controlled by older men. For example, during the Cuban missile crisis between the United States and the Soviet Union in 1961, older men made the decisions that pushed the world to the brink of nuclear war. Thankfully those same men were able to negotiate a settlement in spite of hawkish military leaders such as United States Air Force General Curtis LeMay, who advocated the bombing of Cuba despite the risk of triggering nuclear war. Today, not much has changed. The fingers of old men still hover over the nuclear trigger around the world.

Older men also control many of the important aspects of daily life around the world. A few years ago, I was asked to write an article on the role of men in the future of human evolution in the journal *Futures*.[18] This article became an interesting thought experiment about how male life history traits affect important things like risk tolerance, risky behavior, and perspectives on mortality and reproduction. These questions become more relevant when one considers the fact that mostly men govern our modern lives. All of the top-twenty richest men in the history of the United States have been older men. The United States has always elected a man as president. You get the idea. Despite the many advances in science, politics, and business made by women, it is still a male world. Indeed, I have colleagues who remember when women were not allowed to attend Yale College. They were first admitted in 1969.

My argument in this article was that, since men were not constrained by the same reproductive realities as women, such as childbearing and breastfeeding, men had greater liberty and opportunities to pursue wealth and power. In addition, paternity uncertainty all but surely contributed to the evolution of behavioral traits in men that added to women's family burdens. Consider even today how children with divorced parents almost always live primarily with their mothers. Men also have the opportunity to use that wealth and power to suppress equal opportunities for women. Does this mean that women are biologically predetermined to lack the same opportunities? No. It means that we as a collection of societies that make up our species should be more aware of the need for greater access to child care, family planning, education, and other resources, all factors that contribute mightily to the challenges faced by women.[19]

Most of the world's political and economic power is held and wielded by older men. Given the consistent pattern of male-biased sexual dimorphism in our genus and in earlier hominids, it is likely that males have had more than their share of control over social groups, politics, and reproduction.[20] One could argue that if men are to be the ones to assume positions of power and responsibility, it is much more desirable to have the older guys at the helm since they are less prone than younger men to engage in risky behavior. Better yet, facilitating the empowerment of women and eliminating social, economic, and political barriers would be a tremendously powerful step. If we continue on a path where most of the power and wealth are sequestered with men, it is all but certain that the challenges we face today including the risk of global war resulting in our extinction, climate change, and high-risk economic decisions that have the capacity to devastate entire nations will only continue to grow. Given the narrow margin of error owing to the accelerated pace of global change, it would be wise to continue to amplify the role of women in positions of leadership and influence.

DARWINIAN FEMINISM

In my present role as Deputy Provost for Faculty Development and Diversity, I help Yale find the best and brightest to join our faculty and to make sure our younger professors have every opportunity to generate world-class research and teaching. This is a task that is shared with numerous dedicated faculty and members of the university leadership. To be effective, we need to look at and draw from every available pool of talent. This means scouring the country as well as the rest of the world. Overall, we have been pretty successful. But being good enough is not our goal. We want to be the best, which means continuously raising the bar.

But there is a problem. Historically Yale and other universities have consciously and unconsciously omitted large segments of the population from full consideration to be admitted to the student body as well as the faculty. Those most affected often are women and ethnic/racial minorities.[21] This is unfortunately not unique to Yale; however, we recognize that this omission is antithetical to our goal of identifying the best and brightest. Homogeneity is seldom conducive to scholarly excellence. The opportunity costs are great for Yale, those who have not received full consideration, and society in general. To date, with the exception of Hanna Gray, who served as acting president in 1977, no woman has served as president of Yale since its founding in 1701. Harvard appointed its first woman president several years ago. Much work remains to be done even in these bastions of progressive thought.

Academia is often perceived as being a very progressive and liberal community. Yet even within university faculties, women are paid less, receive unequal consideration for promotion and career advancement, and are generally subject to conscious and unconscious bias when being considered for faculty appointments. Older men make the vast majority of those appointment decisions. Some may dismiss this as politically correct banter. However, the results from numerous scientific studies on unconscious bias against women and underrepresented minorities in science and academia are withering and extremely disconcerting.[22] Our only way forward is to train selection and search committees, presently populated disproportionately by older men, to be aware of these pitfalls and to make sure that the scales are not tipped against those who are on the losing end of these processes.[23] The reason for using this example is that education, economic status, and opportunities for career advancement are all modern sources of variation that *may* influence the evolution of our species. If eliminating sex disparities and reducing older male influence cannot be achieved on the liberal campuses of universities, then we are in big trouble.

Regardless of whether this is relevant to evolution, as long as older men are able to suppress the views and influence of women, our ability to address the problems that our species faces is quite compromised. Opportunities are lost on both sides, by both women and the older men who may be a bit curious about novel ways to move forward on global challenges. Evolutionary biology is not the same as destiny. It simply defines and outlines the constraints and processes by which we change over time. How we manage those constraints and processes will shape how we evolve. In the end, older men need to actively share their power and resources with women. The alternative is to remain mired in the persistent global dilemmas that we face today.

FINAL THOUGHTS

This book started with a discussion of my father, Darwin, Cuategi, and a chimpanzee. The idea was that they all shared the commonality of aging as a male and were the culmination of millions of years of evolution. None of these individuals was perfect. Their reproduction likely accelerated their aging, hormones changed over time and caused them to get rounder around the middle, and more than a few hairs turned gray. But that's ok. Natural selection does not form flawless organisms. It tinkers and muddles through with what was handed down from previous generations. Given that constraint, I think older men and humans in general have done pretty well and have a tremendous capacity to make the world better.

With regard to aging, all men face the prospect of having to adjust to the biophysics of physical deterioration with time and challenges from the environment. Prostates swell, muscle diminishes, younger men challenge you, and naps become ever more attractive. Evolutionary biology can inform us about the origin of age-associated health issues in men and empower us to understand the male aging process with more finesse and depth. Yet despite these challenges, the humanity in older men provided other evolutionary alternatives that made the human condition a bit more tolerable and maybe even improved it, such as the emergence of fathers caring for children, pudgy bellies for cats and grandchildren to snooze on, and perhaps a few more years for everyone to enjoy the wonderful absurdity of life.

NOTES

CHAPTER 1: A GRAY EVOLUTIONARY LENS

1. M. Mangino, "Genomics of Ageing in Twins," *Proceedings of the Nutrition Society* 73 (2014): 526–31, doi:10.1017/S0029665114000640; P. H. Harvey and M. D. Pagel, *The Comparative Method in Evolutionary Biology* (New York: Oxford University Press, 1991). A "baculum" is a penis bone that is quite common in primates and most mammals. For reasons that are unknown, humans do not have one. For an interesting read on the subject, I suggest A. F. Dixson, *Primate Sexuality: Comparative Studies of the Prosimians, Monkeys, Apes, and Humans* (New York: Oxford University Press, 2012), xxii.
2. S. C. Stearns, *The Evolution of Life Histories* (New York: Oxford University Press, 1992).
3. J. Henrich, S. J. Heine, and A. Norenzayan, "The Weirdest People in the World?" *Behavioral and Brain Sciences* 33 (2010): 61–83, discussion 83–135, doi:10.1017/S0140525X0999152X.
4. Throughout this book I use both "hunter-gatherer" and "forager" to describe populations that derive a significant proportion of their food and resources from wild sources. M. R. Rose, *Evolutionary Biology of Aging* (New York: Oxford University Press, 1991).
5. Great apes consist of humans (Homo sapiens), chimpanzees (*Pan sp.*), gorillas (*Gorilla sp.*), and orangutans (*Pongo sp.*). Gibbons and siamangs are considered lesser apes because of their small body size and more distant evolutionary relationships with the great apes.
6. H. Pontzer, D. A. Raichlen, R. W. Shumaker, C. Ocobock, and S. A. Wich, "Metabolic Adaptation for Low Energy Throughput in Orangutans," *Proceedings of the National Academy of Sciences of the United States of America* 107 (2010): 14048–52, doi:10.1073/pnas.1001031107.
7. M. Emery Thompson and C. D. Knott, "Urinary C-peptide of Insulin as a Non-invasive Marker of Energy Balance in Wild Orangutans," *Hormones and Behavior* 53 (2008): 526–35; C. Knott, "Monitoring Health Status of Wild Orangutans through the Field Analysis of Urine" (Poster presented at the American Association of Physical Anthropologists Meeting, Durham, NC, 1996); C. D. Knott, "Changes of Orangutan Caloric Intake, Energy Balance, and Ketones in Response

to Fluctuating Fruit Availability," *International Journal of Primatology* 19 (1998): 1061–79; P. T. Ellison, *On Fertile Ground: A Natural History of Human Reproduction* (Cambridge, MA: Harvard University Press, 2001); G. Jasienska, *The Fragile Wisdom: An Evolutionary View on Women's Biology and Health* (Cambridge, MA: Harvard University Press, 2013); L. L. Sievert, *Menopause: A Biocultural Perspective* (New Brunswick, NJ: Rutgers University Press, 2006).

8. This is based on several elevator rides up Kline Tower on the Yale University campus while serving on a committee with some physicist colleagues.
9. During his travels in Brazil prior to arriving at the Galápagos Islands, Charles Darwin wrote about his disgust for slavery. For example, in his travel memoirs, he notes the following (hardly the kind of statement one would expect from someone who would condone the misuse of his theories in the twentieth century):

> It is often attempted to palliate slavery by comparing the state of slaves with our poorer countrymen: if the misery of our poor be caused not by the laws of nature, but by our institutions, great is our sin; but how this bears on slavery, I cannot see; as well might the use of the thumb-screw be defended in one land, by showing that men in another land suffered from some dreadful disease. Those who look tenderly at the slave owner, and with a cold heart at the slave, never seem to put themselves into the position of the latter;—what a cheerless prospect, with not even a hope of change! picture to yourself the chance, ever hanging over you, of your wife and your little children—those objects which nature urges even the slave to call his own—being torn from you and sold like beasts to the first bidder!

C. Darwin, *The Voyage of the Beagle* (New York: Bantam Books, 1972).

10. S. J. Gould, *The Mismeasure of Man* (New York: W. W. Norton, 1981); J. Marks, *Human Biodiversity: Genes, Race, and History* (Hawthorne, NY: Aldine de Gruyter, 1995); P. Shipman, *The Evolution of Racism: Human Differences and the Use and Abuse of Science* (New York: Simon and Schuster, 1994); P. A. Gowaty, *Feminism and Evolutionary Biology: Boundaries, Intersections, and Frontiers* (New York: Chapman and Hall, 1997); A. Weismann, E. B. Poulton, S. Schönland, and A. E. Shipley, *Essays upon Heredity and Kindred Biological Problems*, 2nd ed. (Oxford: Clarendon Press, 1891); A. Weismann, W. N. Parker, and H. Rōnníeldt, *The Germ-Plasm: A Theory of Heredity* (New York: C. Scribner's Sons, 1893). Compelling evidence has emerged that suggests the existence of transgenerational effects on offspring that seem to result from stresses endured by mothers and even grandmothers. This has significant potential implications for our understanding of health disparities in populations that have experienced historical extreme stresses, such as slavery. See C. E. Aiken and S. E. Ozanne, "Transgenerational Developmental Programming," *Human Reproduction Update* 20, no. 1 (2014): 63–75; D. J. Barker, "The Fetal and Infant Origins of Adult Disease," *BMJ* 301, no. 6761 (1990): 1111; G. Jasienska, "Low Birth Weight of Contemporary African Americans: An Intergenerational Effect of Slavery?" *American Journal of Human Biology* 21, no. 1 (2009): 16–24;

C. W. Kuzawa, "Fetal Origins of Developmental Plasticity: Are Fetal Cues Reliable Predictors of Future Nutritional Environments?" *American Journal of Human Biology* 17, no. 1 (2005): 5–21; and C. Kuzawa and E. Sweet, "Epigenetics and the Embodiment of Race: Developmental Origins of U.S. Racial Disparities in Cardiovascular Health," *American Journal of Human Biology* 21, no. 1 (2009): 2–15. There are some excellent books and articles on the subject of phenotypic plasticity and gene-environment interactions. See A. A. Agrawal, "Phenotypic Plasticity in the Interactions and Evolution of Species," *Science* 294, no. 5541 (2001): 321–26; M. Pigliucci, *Phenotypic Plasticity: Beyond Nature and Nurture* (Baltimore: Johns Hopkins University Press, 2001); S. C. Stearns and J. C. Koella, "The Evolution of Phenotypic Plasticity in Life-History Traits: Predictions of Reaction Norms for Age and Size at Maturity," *Evolution* 40, no. 5 (1986): 893–913; and C. Schlichting and M. Pigliucci, *Phenotypic Evolution: A Reaction Norm Perspective* (Sunderland, MA: Sinauer, 1998).

11. R. M. Henig, *The Monk in the Garden: The Lost and Found Genius of Gregor Mendel, the Father of Genetics* (Boston: Houghton Mifflin, 2000); G. R. Bentley and R. Mace, *Substitute Parents: Biological and Social Perspective on Alloparenting across Human Societies* (New York: Berghahn, 2009); K. L. Kramer, R. D. Greaves, and P. T. Ellison, "Early Reproductive Maturity among Pume Foragers: Implications of a Pooled Energy Model to Fast Life Histories," *American Journal of Human Biology* 21 (2009): 430–37, doi:10.1002/ajhb.20930; K. L. Kramer, "Children's Help and the Pace of Reproduction: Cooperative Breeding in Humans," *Evolutionary Anthropology* 14 (2005): 224–37, doi:10.1002/evan.20082; K. Kramer, *Maya Children: Helpers at the Farm* (Cambridge, MA: Harvard University Press, 2005). For fascinating firsthand accounts of foraging with an Ache band, see the following: K. Hill and K. Hawkes, "Neotropical Hunting among the Ache of Eastern Paraguay," in *Adaptive Responses of Native Amazonians*, ed. R. B. Hames and W. T. Vickers (New York: Academic Press, 1983), 139–88; and K. Hill and A. M. Hurtado, "Hunter-Gatherers of the New World," *American Scientist* 77 (September–October 1989): 437–43. J. Marks, "Without Anthropology, Biological Anthropology Is Just Biology, Only More Poorly Funded," *American Journal of Physical Anthropology* 150 (2013): 189–90.
12. Shipman, *The Evolution of Racism*.

CHAPTER 2: DEAD MAN'S CURVE

1. For those under the age of forty or so, drive-in theaters were massive areas of real estate where people gathered to watch movies on a three-story-high screen from the comfort of their cars. The terrible sound came from heavy metal speakers that hung from your window and clotheslined many a patron going to get popcorn. It is safe to say that those drive-ins made a not so modest contribution to age-specific fertility.

2. Interestingly, the rate of aging can vary depending on the velocity of the individual relative to the observer. But it is only evident or relevant to life if you are traveling very, very, very fast. I strongly suspect that no readers are going to be approaching the speed of light anytime soon so we will shelve that conversation for another day.
3. R. A. Fisher, ed., *The Genetical Theory of Natural Selection* (New York: Dover, 1958), 22–51; T. Dobzhansky, F. J. Ayala, G. L. Stebbins, and J. W. Valentine, *Evolution* (San Francisco: W. H. Freeman, 1977).
4. P. B. Medawar, *An Unsolved Problem of Biology* (Published for the college by H. K. Lewis, 1952).
5. In humans and other multicellular organisms, cells can be roughly categorized into two types, somatic and sexual. Much of an organism is comprised of somatic cells that replicate by mitosis, or the doubling of their genetic material and splitting into two roughly equal daughter cells. This usually occurs to replace worn-out cells or to create new cells during periods of growth. A very small set of special cells called "germ cells" are responsible for creating sex cells that contain half of an organism's genetic complement. These are sperm or ova.
6. R. Skloot, *The Immortal Life of Henrietta Lacks* (New York: Crown, 2010).
7. Some researchers invert these terms, using "senescence" as physical deterioration and "aging" as the passage of time. In this book I use the most common convention.
8. O. Wilde, *The Picture of Dorian Gray* (New York: Illustrated Editions, 1931).
9. R. Pearl, *The Rate of Living, Being an Account of Some Experimental Studies on the Biology of Life Duration* (New York: Knopf, 1928).
10. S. N. Austad, "Comparative Aging and Life Histories in Mammals," *Experimental Gerontology* 32 (1997): 23–38; S. N. Austad and K. E. Fischer, "Mammalian Aging, Metabolism, and Ecology: Evidence from the Bats and Marsupials," *Journal of Gerontology* 46 (1991): B47–53.
11. G. C. Williams, "Pleiotropy, Natural Selection, and the Evolution of Senescence," *Evolution* 11 (1957): 398–411.
12. S. B. Eaton et al., "Women's Reproductive Cancers in Evolutionary Context," *Quarterly Review of Biology* 69 (1994): 353–67.
13. J. K. Kirkwood and M. Rose, "Evolution of Senescence: Late Survival Sacrificed for Reproduction," in *The Evolution of Reproductive Strategies*, ed. L. Southwood (New York: Cambridge University Press, 1991), 15–24.
14. S. N. Austad, "Retarded Senescence in an Insular Population of Virginia Opossums (Didelphis virginiana)," *Journal of Zoology* 229 (1993): 695–708.
15. A. T. Geronimus, "Understanding and Eliminating Racial Inequalities in Women's Health in the United States: The Role of the Weathering Conceptual Framework," *Journal of the American Medical Women's Association* 56 (2001): 133–36, 149–50. This is similar to the idea of "weathering" put forward by sociologist Arline Geronimus to explain chronic health disparities in African American communities. She states, "Weathering suggests that African-American women experience early health deterioration as a consequence of the cumulative impact of repeated experience with social, economic, or political exclusion. This includes the physical cost

of engaging actively to address structural barriers to achievement and well-being" (133).

16. G. Jasienska, I. Nenko, and M. Jasienski, "Daughters Increase Longevity of Fathers, But Daughters and Sons Equally Reduce Longevity of Mothers," *American Journal of Human Biology* 18 (2006): 422–25.
17. A. Ziomkiewicz et al., "Evidence for the Cost of Reproduction in Humans: High Lifetime Reproductive Effort Is Associated with Greater Oxidative Stress in Post-menopausal Women," *PLoS One* 11, no. 1 (2016): e0145753.
18. K. Hill et al., "Mortality Rates among Wild Chimpanzees," *Journal of Human Evolution* 40 (2001): 437–50; K. Hill, A. M. Hurtado, and R. S. Walker, "High Adult Mortality among Hiwi Hunter-Gatherers: Implications for Human Evolution," *Journal of Human Evolution* 52 (2007): 443–54, doi:10.1016/J.Jhevol.2006.11.003; M. N. Muller and R. W. Wrangham, "Mortality Rates among Kanyawara Chimpanzees," *Journal of Human Evolution* 66 (2014): 107–14, doi:http://dx.doi.org/10.1016/j.jhevol.2013.10.004; D. J. Kruger and R. M. Nesse, "An Evolutionary Life-History Framework for Understanding Sex Differences in Human Mortality Rates," *Human Nature* 17 (2006): 74–97; D. J. Kruger and R. M. Nesse, "Sexual Selection and the Male: Female Mortality Ratio," *Evolutionary Psychology* 2 (2004): 66–85; C. O. Lovejoy et al., "Paleodemography of the Libben Site, Ottawa County, Ohio," *Science* 198 (1977): 291–93; K. Hill and A. M. Hurtado, *Ache Life History: The Ecology and Demography of a Foraging People* (Hawthorne, NY: Aldine de Gruyter, 1996); S. N. DeWitte, "Sex Differentials in Frailty in Medieval England," *American Journal of Physical Anthropology* 143 (2010): 285–97, doi:10.1002/ajpa.21316.
19. S. B. Hrdy, "The Optimal Number of Fathers: Evolution, Demography, and History in the Shaping of Female Mate Preferences," *Annals of the New York Academy of Sciences* 907 (2000): 75–96.
20. M. Gurven and H. Kaplan, "Longevity among Hunter-Gatherers: A Cross-Cultural Examination," *Population and Development Review* 33 (2007): 321–65, doi:10.1111/j.1728-4457.2007.00171.x.
21. Hill et al., "Mortality Rates among Wild Chimpanzees."
22. J. R. Carey, *Longevity: The Biology and Demography of Life Span* (Princeton: Princeton University Press, 2003).
23. J. Littleton, "Fifty Years of Chimpanzee Demography at Taronga Park Zoo," *American Journal of Primatology* 67 (2005): 281–98.
24. D. A. Reznick, H. Bryga, and J. A. Endler, "Experimentally Induced Life-History Evolution in a Natural Population," *Nature* 346 (1990): 357–59.
25. In his article, Reznick uses the word "senescence" instead of "aging." This switching of terms occurs occasionally within biodemography and evolutionary biology depending on the preference of the researcher.
26. D. N. Reznick, M. J. Bryant, D. Roff, C. K. Ghalambor, and D. E. Ghalambor, "Effect of Extrinsic Mortality on the Evolution of Senescence in Guppies," *Nature* 431 (2004): 1095–99, doi:nature02936 [pii] 10.1038/nature02936.

27. G. Jasienska, "Reproduction and Lifespan: Trade-offs, Overall Energy Budgets, Intergenerational Costs, and Costs Neglected by Research," *American Journal of Human Biology* 21 (2009): 524–32, doi:10.1002/ajhb.20931.
28. A. A. Maklakov and V. Lummaa, "Evolution of Sex Differences in Lifespan and Aging: Causes and Constraints," *Bioessays* 35 (2013): 717–24, doi:10.1002/bies.201300021; J. C. Regan and L. Partridge, "Gender and Longevity: Why Do Men Die Earlier than Women? Comparative and Experimental Evidence," *Best Practice & Research* 27 (2013): 467–79, doi:10.1016/j.beem.2013.05.016; M. Gurven, H. Kaplan, and A. Z. Supa, "Mortality Experience of Tsimane Amerindians of Bolivia: Regional Variation and Temporal Trends," *American Journal of Human Biology* 19 (2007): 376–98, doi:10.1002/ajhb.20600.
29. R. Rabin, "Health Disparities Persist for Men, and Doctors Ask Why," *New York Times*, November 14, 2006.
30. Regan and Partridge, "Gender and Longevity."
31. T. M. Lee, C. C. Chan, A. W. Leung, P. T. Fox, and J. H. Gao, "Sex-Related Differences in Neural Activity during Risk Taking: An fMRI Study," *Cerebral Cortex* 19 (2009): 1303–12, doi:10.1093/cercor/bhn172.
32. J. Graffelman, E. F. Fugger, K. Keyvanfar, and J. D. Schulman, "Human Live Birth and Sperm-Sex Ratios Compared," *Human Reproduction* 14 (1999): 2917–20.
33. T. Hesketh and Z. W. Xing, "Abnormal Sex Ratios in Human Populations: Causes and Consequences," *Proceedings of the National Academy of Sciences of the United States of America* 103 (2006): 13271–75, doi:10.1073/pnas.0602203103.
34. Gurven, Kaplan, and Supa, "Mortality Experience of Tsimane Amerindians of Bolivia."
35. L. Montville, *Evel: The High-Flying Life of Evel Knievel, American Showman, Daredevil, and Legend*, 1st ed. (New York: Doubleday, 2011).
36. M. Daly and M. Wilson, *Sex, Evolution, and Behavior*, 2nd ed. (Belmont, CA: Wadsworth Publishing, 1983).
37. Kruger and Nesse, "Sexual Selection and the Male."
38. B. Bogin, *Patterns of Human Growth*, 2nd ed. (New York: Cambridge University Press, 1999).
39. J. R. Goldstein, "A Secular Trend toward Earlier Male Sexual Maturity: Evidence from Shifting Ages of Male Young Adult Mortality," *PLoS One* 6, no. 8 (2011): e14826, doi:10.1371/journal.pone.0014826.
40. M. Oakwood, A. J. Bradley, and A. Cockburn, "Semelparity in a Large Marsupial," *Proceedings: Biological Sciences* 268 (2001): 407–11.
41. S. W. Gangestad, J. A. Simpson, A. J. Cousins, C. E. Garver-Apgar, and P. N. Christensen, "Women's Preferences for Male Behavioral Displays Change across the Menstrual Cycle," *Psychological Science* 15 (2004): 203–7, doi:10.1111/j.0956-7976.2004.01503010.x.
42. J. R. Roney and Z. L. Simmons, "Women's Estradiol Predicts Preference for Facial Cues of Men's Testosterone," *Hormones and Behavior* 53 (2008): 14–19, doi:10.1016/j.yhbeh.2007.09.008.

43. Kruger and Nesse, "Sexual Selection and the Male"; T. H. Clutton-Brock and G. R. Iason, "Sex Ratio Variation in Mammals," *Quarterly Review of Biology* 61 (1986): 339–74.
44. L. Vinicius, R. Mace, and A. Migliano, "Variation in Male Reproductive Longevity across Traditional Societies," *PLoS One* 9 (2014): e112236, doi:10.1371/journal.pone.0112236.
45. M. W. Reiches et al., "Pooled Energy Budget and Human Life History," *American Journal of Human Biology* 21 (2009): 421–29, doi:10.1002/ajhb.20906.
46. H. Kaplan, K. Hill, J. Lancaster, and A. M. Hurtado, "A Theory of Human Life History Evolution: Diet, Intelligence and Longevity," *Evolutionary Anthropology* 9 (2000): 156–84.
47. K. R. Hill et al., "Co-residence Patterns in Hunter-Gatherer Societies Show Unique Human Social Structure," *Science* 331 (2011): 1286–89, doi:10.1126/science.1199071.
48. A. Bonmati et al., "Middle Pleistocene Lower Back and Pelvis from an Aged Human Individual from the Sima de los Huesos Site, Spain," *Proceedings of the National Academy of Sciences of the United States of America* 107 (2010): 18386–91, doi:10.1073/pnas.1012131107.
49. Jasienska, Nenko, and Jasienski, "Daughters Increase Longevity of Fathers."

CHAPTER 3: GETTING A HANDLE ON LOVE HANDLES

1. C. Darwin and D. King-Hele, *The Life of Erasmus Darwin,* 1st unabridged ed. (New York: Cambridge University Press, 2003); P. Fara, *Erasmus Darwin: Sex, Science, and Serendipity* (New York: Oxford University Press, 2012).
2. N. T. Roach, M. Venkadesan, M. J. Rainbow, and D. E. Lieberman, "Elastic Energy Storage in the Shoulder and the Evolution of High-Speed Throwing in Homo," *Nature* 498 (2013): 483–86, doi:10.1038/nature12267.
3. M. Elia, "Organ and Tissue Contribution to Metabolic Rate," in *Energy Metabolism: Tissue Determinants and Cellular Corollaries,* ed. J. M. Kinney and H. N. Tucker (New York: Raven Press, 1992), 61–79.
4. J. Henriksson, "Energy Metabolism in Muscle: Its Possible Role in the Adaptation to Energy Deficiency," in *Energy Metabolism: Tissue Determinants and Cellular Corollaries,* ed. J. M. Kinney and H. N. Tucker (New York: Raven Press, 1992), 345–65.
5. K. Sato, M. Iemitsu, K. Aizawa, and R. Ajisaka, "Testosterone and DHEA Activate the Glucose Metabolism-Related Signaling Pathway in Skeletal Muscle," *American Journal of Physiology-Endocrinology and Metabolism* 294 (2008): E961–68, doi:10.1152/ajpendo.00678.2007.
6. P. T. Ellison et al., "Population Variation in Age-Related Decline in Male Salivary Testosterone," *Human Reproduction* 17 (2002): 3251–53.

7. A. Uchida et al., "Age Related Variation of Salivary Testosterone Values in Healthy Japanese Males," *Aging Male* 9 (2006): 207–13.
8. A. Gray, H. A. Feldman, J. B. McKinlay, and C. Longcope, "Age, Disease, and Changing Sex Hormone Levels in Middle-Aged Men: Results of the Massachusetts Male Aging Study," *Journal of Clinical Endocrinology and Metabolism* 73 (1991): 1016–25.
9. D. Joshi et al., "Low Free Testosterone Levels Are Associated with Prevalence and Incidence of Depressive Symptoms in Older Men," *Clinical Endocrinology (Oxford)* 72 (2010): 232–40, doi:10.1111/j.1365-2265.2009.03641.x; E. J. Giltay et al., "Salivary Testosterone: Associations with Depression, Anxiety Disorders, and Antidepressant Use in a Large Cohort Study," *Journal of Psychosomatic Research* 72 (2012): 205–13, doi:10.1016/j.jpsychores.2011.11.014.
10. F. R. Sattler et al., "Testosterone and Growth Hormone Improve Body Composition and Muscle Performance in Older Men," *Journal of Clinical Endocrinology and Metabolism* 94 (2009): 1991–2001, doi:10.1210/jc.2008-2338.
11. B. C. Trumble et al., "Age-Independent Increases in Male Salivary Testosterone during Horticultural Activity among Tsimane Forager-Farmers," *Evolution and Human Behavior* 34 (2013), doi:10.1016/j.evolhumbehav.2013.06.002.
12. W. J. Bremner, M. V. Vitiello, and P. N. Prinz, "Loss of Circadian Rhythmicity in Blood Testosterone Levels with Aging in Normal Men," *Journal of Clinical Endocrinology and Metabolism* 56 (1983): 1278–81; R. G. Bribiescas and K. R. Hill, "Circadian Variation in Salivary Testosterone across Age Classes in Ache Amerindian Males of Paraguay," *American Journal of Human Biology* 22 (2010): 216–20, doi:10.1002/ajhb.21012.
13. D. Amir, P. T. Ellison, K. R. Hill, and R. G. Bribiescas, "Diurnal Variation in Salivary Cortisol across Age Classes in Ache Amerindian Males of Paraguay," *American Journal of Human Biology* 27 (2015): 344–48, doi:10.1002/ajhb.22645.
14. Jasienska, "Reproduction and Lifespan."
15. J. C. Brüning et al., "Role of Brain Insulin Receptor in Control of Body Weight and Reproduction," *Science* 289 (2000): 2122–25.
16. S. Röjdmark, "Influence of Short-Term Fasting on the Pituitary-Testicular Axis in Normal Men," *Hormone Research* 25 (1987): 140–46.
17. S. Röjdmark, "Increased Gonadotropin Responsiveness to Gonadotropin-Releasing Hormone during Fasting in Normal Subjects," *Metabolism* 36 (1987): 21–26.
18. The hypothalamus is located at the core base of the brain. It is a tiny collection of neurons that regulate many basic functions including hunger, thirst, and reproduction. The hypothalamus is connected to the pituitary, a small pebble-sized gland at the front base of the brain. It receives hormonal signals from the hypothalamus by way of a vessel not much wider than a human hair. The pituitary is stimulated by GnRH and other similar hormones to produce other hormones that regulate reproduction, metabolism, and a number of other endocrine agents.

19. J. D. Veldhuis et al., "Amplitude Suppression of the Pulsatile Mode of Immunoradiometric Luteinizing Hormone Release in Fasting-Induced Hypoandrogenemia in Normal Men," *Journal of Clinical Endocrinology and Metabolism* 76 (1993): 587–93.
20. P. A. Nepomnaschy, K. Welch, D. McConnell, B. I. Strassmann, and B. G. England, "Stress and Female Reproductive Function: A Study of Daily Variations in Cortisol, Gonadotrophins, and Gonadal Steroids in a Rural Mayan Population," *American Journal of Human Biology* 16 (2004): 523–32; P. A. Nepomnaschy et al., "Cortisol Levels and Very Early Pregnancy Loss in Humans," *Proceedings of the National Academy of Sciences of the United States of America* 103 (2006): 3938–42, doi:0511183103 [pii], 10.1073/pnas.0511183103; C. J. Hobel, C. Dunkel-Schetter, S. C. Roesch, L. C. Castro, and C. P. Arora, "Maternal Plasma Corticotropin-Releasing Hormone Associated with Stress at 20 Weeks' Gestation in Pregnancies Ending in Preterm Delivery," *American Journal of Obstetrics & Gynecology* 180 (1999): S257–S263, doi:10.1016/S0002-9378(99)70712-X.
21. R. M. Tamimi et al., "Average Energy Intake among Pregnant Women Carrying a Boy Compared with a Girl," *British Medical Journal* 326 (2003): 1245–46.
22. J. M. Tanner, *Fetus into Man: Physical Growth from Conception to Maturity* (Cambridge, MA: Harvard University Press, 1978); J. C. Wells, "Sexual Dimorphism of Body Composition," *Best Practice & Research* 21 (2007): 415–30, doi:10.1016/j.beem.2007.04.007; J. M. Kinney and H. N. Tucker, eds., *Energy Metabolism: Tissue Determinants and Cellular Corollaries* (New York: Raven Press, 1992).
23. Elia, "Organ and Tissue Contribution to Metabolic Rate," 61–79.
24. N. K. Fukagawa, L. G. Bandini, and J. B. Young, "Effect of Age on Body Composition and Resting Metabolic Rate," *American Journal of Physiology* 259 (1990): E233–238.
25. One may wonder why the sample sizes are so small. The reason is twofold. First, Hadza communities tend to be small, and it is very challenging to conduct this type of research owing to their geographic remoteness. Second, single doses of doubly labeled water are extremely expensive; an espresso-sized dose costs around $500.
26. Basal metabolic rates from the Shuar of Ecuador using doubly labeled water have yielded similar results, although the entire analysis has not yet been peer reviewed or published. H. Pontzer et al., "Hunter-Gatherer Energetics and Human Obesity," *PLoS One* 7 (2012): e40503, doi:10.1371/journal.pone.0040503; L. Christopher, F. C. Madimenos, R. G. Bribiescas, L. S. Sugiyama, and H. Pontzer, "Daily Energy Requirements of Shuar Forager-Horticulturalists," *American Journal of Physical Anthropology* (in press).
27. R. G. Bribiescas, "Testosterone Levels among Aché Hunter-Gatherer Men: A Functional Interpretation of Population Variation among Adult Males," *Human Nature* 7 (1996): 163–88; R. G. Bribiescas, "Serum Leptin Levels and Anthropometric Correlates in Ache Amerindians of Eastern Paraguay," *American Journal of Physical Anthropology* 115 (2001): 297–303; P. T. Ellison, "Energetics and Reproductive Effort," *American Journal of Human Biology* 15 (2003): 342–51.

28. C. A. Marler and M. C. Moore, "Evolutionary Costs of Aggression Revealed by Testosterone Manipulations in Free-Living Lizards," *Behavioral Ecology and Sociobiology* 23 (1988): 21–26; C. A. Marler, G. Walsberg, M. L. White, and M. Moore, "Increased Energy Expenditure Due to the Increased Territorial Defense in Male Lizards after Phenotypic Manipulation," *Behavioral Ecology and Sociobiology* 37 (1995): 225–31; W. L. Reed et al., "Physiological Effects on Demography: A Long-Term Experimental Study of Testosterone's Effects on Fitness," *American Naturalist* 167 (2006): 665–81.
29. Elia, "Organ and Tissue Contribution to Metabolic Rate."
30. A. A. Khazaeli, W. Van Voorhies, and J. W. Curtsinger, "Longevity and Metabolism in Drosophila Melanogaster: Genetic Correlations between Life Span and Age-Specific Metabolic Rate in Populations Artificially Selected for Long Life," *Genetics* 169 (2005): 231–42, doi:10.1534/genetics.104.030403.
31. W. Shen et al., "Sexual Dimorphism of Adipose Tissue Distribution across the Lifespan: A Cross-Sectional Whole-Body Magnetic Resonance Imaging Study," *Nutrition & Metabolism* 6 (2009): 17, doi:10.1186/1743-7075-6-17.
32. B. M. Filippi and T. K. Lam, "Leptin and Aging," *Aging (Albany NY)* 6 (2014): 82–83; M. Balasko, S. Soos, M. Szekely, and E. Petervari. "Leptin and Aging: Review and Questions with Particular Emphasis on Its Role in the Central Regulation of Energy Balance," *Journal of Chemical Neuroanatomy* 61–62 (2014): 248–55, doi:10.1016/j.jchemneu.2014.08.006.
33. M. Abdelgadir, M. Elbagir, M. Eltom, C. Berne, and B. Ahren, "Reduced Leptin Concentrations in Subjects with Type 2 Diabetes Mellitus in Sudan," *Metabolism* 51 (2002): 304–6; Bribiescas, "Serum Leptin Levels and Anthropometric Correlates in Ache Amerindians of Eastern Paraguay"; C. Fox et al., "Plasma Leptin Concentrations in Pima Indians Living in Drastically Different Environments," *Diabetes Care* 22 (1999): 413–17; S. Lindeberg, S. Soderberg, B. Ahren, and T. Olsson, "Large Differences in Serum Leptin Levels between Nonwesternized and Westernized Populations: The Kitava Study," *Journal of Internal Medicine* 249 (2001): 553–58; F. Lindgarde, M. B. Ercilla, L. R. Correa, and B. Ahren, "Body Adiposity, Insulin, and Leptin in Subgroups of Peruvian Amerindians," *High Altitude Medicine & Biology* 5 (2004): 27–31; F. Lindgarde, I. Widen, M. Gebb, and B. Ahren, "Traditional versus Agricultural Lfestyle among Shuar Women of the Ecuadorian Amazon: Effects on Leptin Levels," *Metabolism* 53 (2004): 1355–58.
34. R. G. Bribiescas and S. F. Anestis, "Leptin Associations with Age, Weight, and Sex among Chimpanzees (Pan troglodytes)," *Journal of Medical Primatology* 39 (2010): 347–55, doi:JMP419 [pii], 10.1111/j.1600-0684.2010.00419.x; R. G. Bribiescas and M. S. Hickey, "Population Variation and Differences in Serum Leptin Independent of Adiposity: A Comparison of Ache Amerindian Men of Paraguay and Lean American Male Distance Runners," *Nutrition & Metabolism* 3 (2006): 34.
35. I. Czajka-Oraniec, W. Zgliczynski, A. Kurylowicz, M. Mikula, and J. Ostrowski, "Association between Gynecomastia and Aromatase (CYP19) Polymorphisms," *European Journal of Endocrinology* 158 (2008): 721–27, doi:10.1530/EJE-07-0556.

36. D. M. Schulte et al., "Caloric Restriction Increases Serum Testosterone Concentrations in Obese Male Subjects by Two Distinct Mechanisms," *Hormone and Metabolic Research* 46 (2014): 283–86, doi:10.1055/s-0033-1358678.
37. R. G. Bribiescas, "Age-Related Differences in Serum Gonadotropin (FSH and LH), Salivary Testosterone, and 17-Beta Estradiol Levels among Ache Amerindian Males of Paraguay," *American Journal of Physical Anthropology* 127 (2005): 114–21.
38. K. B. Clancy, "Reproductive Ecology and the Endometrium: Physiology, Variation, and New Directions," *American Journal of Physical Anthropology* 140 Suppl. 49 (2009): 137–54, doi:10.1002/ajpa.21188; J. V. Durnin, "Energy Requirements of Pregnancy," *Acta Paediatrica Scandinavica* 373 (1991): 33–42; A. M. Prentice and A. Prentice, "Energy Costs of Lactation," *Annual Review of Nutrition* 8 (1988): 63–79.
39. C. Key and C. Ross, "Sex Differences in Energy Expenditure in Non-human Primates," *Proceedings: Biological Sciences/The Royal Society* 266 (1999): 2479–85.
40. Bribiescas, "Serum Leptin Levels and Anthropometric Correlates in Ache Amerindians of Eastern Paraguay."
41. C. Ruggiero et al., "High Basal Metabolic Rate Is a Risk Factor for Mortality: The Baltimore Longitudinal Study of Aging," *Journals of Gerontology: Series A, Biological Sciences and Medical Sciences* 63 (2008): 698–706.
42. M. P. Muehlenbein, J. L. Hirschtick, J. Z. Bonner, and A. M. Swartz, "Toward Quantifying the Usage Costs of Human Immunity: Altered Metabolic Rates and Hormone Levels during Acute Immune Activation in Men," *American Journal of Human Biology* 22 (2010): 546–56, doi:10.1002/ajhb.21045.
43. S. M. Garn and L. C. Clark, Jr., "The Sex Difference in the Basal Metabolic Rate," *Child Development* 24 (1953): 215–24.
44. R. Adelman, R. L. Saul, and B. N. Ames, "Oxidative Damage to DNA: Relation to Species Metabolic Rate and Life Span," *Proceedings of the National Academy of Sciences of the United States of America* 85 (1988): 2706–8.
45. G. Block et al., "Factors Associated with Oxidative Stress in Human Populations," *American Journal of Epidemiology* 156 (2002): 274–85, doi:10.1093/aje/kwf029.
46. S. Loft et al., "Oxidative DNA Damage Estimated by 8-Hydroxydeoxyguanosine Excretion in Humans: Influence of Smoking, Gender and Body Mass Index," *Carcinogenesis* 13 (1992): 2241–47.
47. C. Alonso-Alvarez, S. Bertrand, B. Faivre, O. Chastel, and G. Sorci, "Testosterone and Oxidative Stress: The Oxidation Handicap Hypothesis," *Proceedings of the Royal Society B: Biological Sciences* 274 (2007): 819–25, doi:10.1098/rspb.2006.3764.
48. Jasienska, "Reproduction and Lifespan."
49. Ziomkiewicz et al., "Evidence for the Cost of Reproduction in Humans."
50. J. M. Plavcan, "Sexual Dimorphism in Primate Evolution," *American Journal of Physical Anthropology* Suppl. 33 (2001): 25–53.
51. R. G. Bribiescas, "Testosterone as a Proximate Determinant of Somatic Energy Allocation in Human Males: Evidence from Ache Men of Eastern Paraguay" (PhD diss., Harvard University, 1997).

52. R. Walker and K. Hill, "Modeling Growth and Senescence in Physical Performance among the Ache of Eastern Paraguay," *American Journal of Human Biology* 15 (2003): 196–208.
53. A. C. Pisor, M. Gurven, A. D. Blackwell, H. Kaplan, and G. Yetish, "Patterns of Senescence in Human Cardiovascular Fitness: VO2 Max in Subsistence and Industrialized Populations," *American Journal of Human Biology* 25 (2013): 756–69, doi:10.1002/ajhb.22445.
54. R. Walker, K. Hill, H. Kaplan, and G. McMillan, "Age-Dependency in Hunting Ability among the Ache of Eastern Paraguay," *Journal of Human Evolution* 42, no. 6 (2002): 639–57.
55. Kaplan et al., "A Theory of Human Life History Evolution."

CHAPTER 4: OLDER FATHERS, LONGER LIVES

1. M. L. Walker and J. G. Herndon, "Menopause in Nonhuman Primates?" *Biological Reproduction* 79 (2008): 398–406, doi:10.1095/biolreprod.108.068536.
2. Julie Meyer, "Centenarians: 2010," 2010 Census Special Reports, U.S. Census Bureau, December 13 2012, http://www.census.gov/prod/cen2010/reports/c2010sr-03.pdf%3E.
3. "Life Expectancy at Birth, Total (Years)," World Bank, n.d., http://data.worldbank.org/indicator/SP.DYN.LE00.IN%3E.
4. Organismic characteristics that affect life span are metabolism, body size, and types of growth. Reptiles are ectotherms; that is, they derive their body heat from the surrounding environment and therefore have slower metabolisms. This contributes to the physiology of aging and life span. This is the reason large snakes such as pythons and anacondas reach such large sizes and live a long time. The same applies to tortoises. Birds are a bit of an enigma since they are small bodied and have fast metabolisms but relatively long lives. Think of parrots and macaws that live several decades. This is the reason we are restricting our comparisions of human life span to other great apes.
5. Bogin, *Patterns of Human Growth.*
6. K. K. Catlett, G. T. Schwartz, L. R. Godfrey, and W. L. Jungers, "'Life History Space': A Multivariate Analysis of Life History Variation in Extant and Extinct Malagasy Lemurs," *American Journal of Physical Anthropology* 142 (2010): 391–404, doi:10.1002/ajpa.21236; S. Atsalis and S. Margulis, "Primate Reproductive Aging: From Lemurs to Humans," *Interdisciplinary Topics in Gerontology* 36 (2008): 186–94, doi:10.1159/000137710; P. Wright, S. J. King, A. Baden, and J. Jernvall, "Aging in Wild Female Lemurs: Sustained Fertility with Increased Infant Mortality," *Interdisciplinary Topics in Gerontology* 36 (2008): 17–28, doi:10.1159/000137677; A. F. Richard, R. E. Dewar, M. Schwartz, and L. Ratsirarson, "Life in the Slow Lane? Demography and Life Histories of Male and Female Sifaka (Propithecus verreauxi verreauxi)," *Journal of Zoology, London* 256 (2002): 421–36.

7. Austad, "Comparative Aging and Life Histories in Mammals"; S. N. Austad and K. E. Fischer, "Primate Longevity: Its Place in the Mammalian Scheme," *American Journal of Primatology* 28 (1992): 251–61.
8. J. H. Jones, "Primates and the Evolution of Long, Slow Life Histories," *Current Biology* 21 (2011): R708–717, doi:10.1016/j.cub.2011.08.025.
9. O. Burger, R. Walker, and M. J. Hamilton, "Lifetime Reproductive Effort in Humans," *Proceedings of the Royal Society B: Biological Sciences* 277 (2010): 773–77, doi:10.1098/rspb.2009.1450.
10. R. W. Wrangham and D. Peterson, *Demonic Males: Apes and the Origins of Human Violence* (Boston: Houghton Mifflin, 1996).
11. B. Kuhnert and E. Nieschlag, "Reproductive Functions of the Ageing Male," *Human Reproduction Update* 10, no. 4 (2004): 327–39.
12. A. J. Wilcox et al., "Incidence of Early Loss of Pregnancy," *New England Journal of Medicine* 319 (1988): 189–94.
13. K. G. Anderson, "How Well Does Paternity Confidence Match Actual Paternity? Evidence from Worldwide Nonpaternity Rates," *Current Anthropology* 47 (2006): 513–20, doi:10.1086/504167.
14. E. de La Rochebrochard, J. de Mouzon, F. Thepot, and P. Thonneau, "Fathers over 40 and Increased Failure to Conceive: The Lessons of In Vitro Fertilization in France," *Fertility and Sterility* 85 (2006): 1420–24.
15. Jasienska, Nenko, and Jasienski, "Daughters Increase Longevity of Fathers."
16. M. L. Eisenberg et al., "Semen Quality, Infertility and Mortality in the USA," *Human Reproduction* 29, no. 7 (2014): 1567–74, doi:10.1093/humrep/deu106.
17. H. I. Kort et al., "Impact of Body Mass Index Values on Sperm Quantity and Quality," *Journal of Andrology* 27 (2006): 450–52, doi:jandrol.05124 [pii], 10.2164/jandrol.05124; E. M. Pauli et al., "Diminished Paternity and Gonadal Function with Increasing Obesity in Men," *Fertility and Sterility* 90 (2008): 346–51.
18. Walker and Herndon, "Menopause in Nonhuman Primates?"
19. Hill and Hurtado, *Ache Life History.*
20. K. P. Jones et al., "Depletion of Ovarian Follicles with Age in Chimpanzees: Similarities to Humans," *Biology of Reproduction* 77 (2007): 247–51.
21. M. Emery Thompson et al., "Aging and Fertility Patterns in Wild Chimpanzees Provide Insights into the Evolution of Menopause," *Current Biology* 17, no. 24 (2007): 2150–56.
22. Ellison, *On Fertile Ground.*
23. P. S. Kim, J. E. Coxworth, and K. Hawkes, "Increased Longevity Evolves from Grandmothering," *Proceedings: Biological Sciences/The Royal Society* 279 (2012): 4880–84, doi:10.1098/rspb.2012.1751.
24. I. Perez-Alcala, L. L. Sievert, C. M. Obermeyer, and D. S. Reher, "Cross Cultural Analysis of Factors Associated with Age at Natural Menopause among Latin-American Immigrants to Madrid and Their Spanish Neighbors," *American Journal of Human Biology* 25 (2013): 780–88, doi:10.1002/ajhb.22447.

25. Jones et al., "Depletion of Ovarian Follicles with Age in Chimpanzees"; Emery Thompson et al., "Aging and Fertility Patterns in Wild Chimpanzees."
26. F. Marlowe, "Male Care and Mating Effort among Hadza Foragers," *Behavioral Ecology and Sociobiology* 46 (1999): 46–57.
27. F. Marlowe, "The Patriarch Hypothesis: An Alternative Explanation of Menopause," *Human Nature* 11 (2000): 27–42.
28. Vinicius, Mace, and Migliano, "Variation in Male Reproductive Longevity across Traditional Societies."
29. H. Kaplan and K. Hill, "Hunting Ability and Reproductive Success among Male Ache Foragers: Preliminary Results," *Current Anthropology* 26 (1985): 131–33.
30. De La Rochebrochard et al., "Fathers over 40 and Increased Failure to Conceive."
31. Bribiescas, "Age-Related Differences in Serum Gonadotropin (FSH and LH), Salivary Testosterone, and 17-Beta Estradiol Levels among Ache Amerindian Males of Paraguay."
32. S. D. Tuljapurkar, C. O. Puleston, and M. D. Gurven, "Why Men Matter: Mating Patterns Drive Evolution of Human Lifespan," *PLoS One* 2 (2007): e785, doi:10.1371/journal.pone.0000785.

CHAPTER 5: DEAR OLD DAD

1. M. C. Inhorn, W. Chavkin, and J.-A. Navarro, *Globalized Fatherhood* (New York: Berghahn, 2015).
2. B. S. Hewlett, ed., *Father-Child Relations: Cultural and Biosocial Contexts* (Hawthorne, NY: Aldine de Gruyter, 1992); P. B. Gray and K. G. Anderson, *Fatherhood: Evolution and Human Paternal Behavior* (Cambridge, MA: Harvard University Press, 2010).
3. M. V. Flinn, "Paternal Care in a Caribbean Village," in *Father-Child Relations: Cultural and Biosocial Contexts*, ed. B. S. Hewlett (Hawthorne, NY: Aldine de Gruyter, 1992), 57–84; J. Stieglitz, M. Gurven, H. Kaplan, and J. Winking, "Infidelity, Jealousy, and Wife Abuse among Tsimane Forager-Farmers: Testing Evolutionary Hypotheses of Marital Conflict," *Evolution and Human Behavior* 33 (2012): 438–48, doi:10.1016/j.evolhumbehav.2011.12.006; J. Stieglitz, H. Kaplan, M. Gurven, J. Winking, and B. V. Tayo, "Spousal Violence and Paternal Disinvestment among Tsimane' Forager-Horticulturalists," *American Journal of Human Biology* 23 (2011): 445–57, doi:10.1002/ajhb.21149.
4. Some of you may be saying, "What about sea horses?" First, they aren't mammals. Second, yes, they are a rare exception. Males do not commonly gestate offspring.
5. J. E. Mank, D. E. Promislow, and J. C. Avise, "Phylogenetic Perspectives in the Evolution of Parental Care in Ray-Finned Fishes," *Evolution* 59 (2005): 1570–78.
6. K. Hawkes, "Human Longevity: The Grandmother Effect," *Nature* 428 (2004): 128–29; K. Hawkes, J. F. O'Connell, N. G. B. Jones, H. Alvarez, and E. L. Charnov, "Grandmothering, Menopause, and the Evolution of Human Life Histories," *Proceedings of the National Academy of Sciences* 95 (1998): 1336–39.

7. Kruger and Nesse, "An Evolutionary Life-History Framework for Understanding Sex Differences in Human Mortality Rates."
8. T. C. Burnham et al., "Men in Committed, Romantic Relationships Have Lower Testosterone," *Hormones and Behavior* 44 (2003): 119–22.
9. E. Fernandez-Duque, C. R. Valeggia, and S. P. Mendoza, "The Biology of Paternal Care in Human and Nonhuman Primates," *Annual Review of Anthropology* 38 (2009): 115–30, doi:10.1146/annurev-anthro-091908-164334.
10. P. B. Gray and J. R. Garcia, *Evolution and Human Sexual Behavior* (Cambridge, MA: Harvard University Press, 2013).
11. A. V. Jaeggi, B. C. Trumble, H. S. Kaplan, and M. Gurven, "Salivary Oxytocin Increases Concurrently with Testosterone and Time Away from Home among Returning Tsimane' Hunters," *Biology Letters* 11 (2015), doi:10.1098/rsbl.2015.0058.
12. Hrdy, "The Optimal Number of Fathers." Recall that if a male mates with a hundred females in a year, he could father a hundred offspring. The same is not true for females. If she mates with a hundred males, she will not have a hundred offspring. While there are many potential benefits for females to have many mates, increased fertility on par with males is not one of them.
13. A. M. Hurtado and K. R. Hill, "Paternal Effect on Offspring Survivorship among Ache and Hiwi Hunter-Gatherers: Implications for Modeling Pair-Bond Stability," in *Father-Child Relations: Cultural and Biosocial Contexts*, ed. B. S. Hewlett (Hawthorne, NY: Aldine de Gruyter, 1992), 31–56.
14. Ibid.
15. Kaplan and Hill, "Hunting Ability and Reproductive Success among Male Ache Foragers."
16. W. Saltzman and D. Maestripieri, "The Neuroendocrinology of Primate Maternal Behavior," *Progress in Neuro-Psychopharmacology & Biological Psychiatry* 35 (2011): 1192–1204, doi:10.1016/j.pnpbp.2010.09.017.
17. E. D. Ketterson et al., "Testosterone, Phenotype and Fitness: A Research Program in Evolutionary Behavioral Endocrinology," in *Avian Endocrinology*, ed. C. M. Chaturvedi (New Delhi: Narosa Publishing House, 2001), 19–40; E. D. Ketterson, V. Nolan Jr., L. Wolf, and C. Ziegenfus, "Testosterone and Avian Life Histories: Effects of Experimentally Elevated Testosterone on Behavior and Correlates of Fitness in the Dark-Eyed Junco (Junco hyemalis)," *American Naturalist* 140 (1992): 980–99; S. J. Schoech, E. D. Ketterson, V. Nolan Jr., P. J. Sharp, and J. D. Buntin, "The Effect of Exogenous Testosterone on Parental Behavior, Plasma Prolactin, and Prolactin Binding Sites in Dark-Eyed Juncos," *Hormones and Behavior* 34 (1998): 1–10.
18. V. S. Gromov and L. V. Osadchuk, "Parental Care and Testosterone in Males of the Bank Vole (Myodes glareolus): Sensitization and Androgenic Stimulation of Paternal Behavior," *Biology Bulletin of the Russian Academy of Sciences* 40 (2013): 114–18, doi:10.1134/S1062359013010056; V. S. Gromov and V. V. Voznesenskaya, "Care of Young, Aggressiveness, and Secretion of Testosterone in Male Rodents: A Correlation Analysis," *Biology Bulletin of the Russian Academy of Sciences* 40 (2013): 463–70, doi:10.1134/S1062359013050063.

19. Gray and Anderson, *Fatherhood*; P. B. Gray, "Marriage, Parenting, and Testosterone Variation among Kenyan Swahili Men," *American Journal of Physical Anthropology* 122 (2003): 279–86; P. B. Gray, S. M. Kahlenberg, E. S. Barrett, S. F. Lipson, and P. T. Ellison, "Marriage and Fatherhood Are Associated with Lower Testosterone in Males," *Evolution and Human Behavior* 23 (2002): 193–201; P. B. Gray, J. C. Parkin, and M. E. Samms-Vaughan, "Hormonal Correlates of Human Paternal Interactions: A Hospital-Based Investigation in Urban Jamaica," *Hormones and Behavior* 52 (2007): 499–507, doi:S0018-506X(07)00167-5 [pii], 10.1016/j.yhbeh.2007.07.005; P. B. Gray, C. F. Yang, and H. G. Pope Jr., "Fathers Have Lower Salivary Testosterone Levels than Unmarried Men and Married Non-fathers in Beijing, China," *Proceedings: Biological Sciences/The Royal Society* 273 (2006): 333–39.
20. L. T. Gettler, T. W. McDade, A. B. Feranil, and C. W. Kuzawa, "Longitudinal Evidence That Fatherhood Decreases Testosterone in Human Males," *Proceedings of the National Academy of Sciences* 108, no. 39 (2011): 16194–99, doi:10.1073/pnas.1105403108.
21. M. N. Muller, F. W. Marlow, R. Bugumba, and P. T. Ellison, "Testosterone and Paternal Care in East African Foragers and Pastoralists," *Proceedings of the Royal Society, Biological Sciences* 276 (2009): 347–54.
22. Ibid.
23. Interestingly, women who are married or living as married exhibit higher levels of progesterone and estradiol compared to unmarried or paired women after controlling for energetics and other relevant factors. Clearly reproductive hormones are central to reproductive states and bonding in both sexes. See E. S. Barrett, V. Tran, S. W. Thurston, H. Frydenberg, S. F. Lipson, I. Thune, and P. T. Ellison, "Women Who Are Married or Living as Married Have Higher Salivary Estradiol and Progesterone than Unmarried Women," *American Journal of Human Biology* 27, no. 4 (2015): 501–7.
24. A. Goldberg, "Evaluating the Relationship between Fatherhood and Immunocompetence" (B.A. thesis, Yale University, 2008); M. P. Muehlenbein and R. G. Bribiescas, "Testosterone-Mediated Immune Functions and Male Life Histories," *American Journal of Human Biology* 17 (2005): 527–58.
25. J. C. K. Wells, *The Evolutionary Biology of Human Body Fatness: Thrift and Control* (New York: Cambridge University Press, 2010).
26. E. R. Hofny et al., "Semen Parameters and Hormonal Profile in Obese Fertile and Infertile Males," *Fertility and Sterility* 94, no. 2 (2009): 581–84, doi:S0015-0282(09)00755-9 [pii], 10.1016/j.fertnstert.2009.03.085.
27. Wells, *The Evolutionary Biology of Human Body Fatness.*
28. M. J. Rantala et al., "Adiposity, Compared with Masculinity, Serves as a More Valid Cue to Immunocompetence in Human Mate Choice," *Proceedings of the Royal Society, Biological Sciences* 280 (2013): 20122495, doi:10.1098/rspb.2012.2495.
29. Bribiescas and Anestis, "Leptin Associations with Age, Weight, and Sex among Chimpanzees."
30. C. F. Garfield et al., "Longitudinal Study of Body Mass Index in Young Males and the Transition to Fatherhood," *American Journal of Men's Health* (2015), doi:10.1177/1557988315596224.

31. G. Hausfater and S. B. Hrdy, eds., *Infanticide: Comparative and Evolutionary Perspectives* (New York: Aldine Publishing, 1984).
32. Gray and Anderson, *Fatherhood*; R. G. Bribiescas, P. T. Ellison, and P. B. Gray, "Male Life History, Reproductive Effort, and the Evolution of the Genus Homo: New Directions and Perspectives," *Current Anthropology* 55 (2012): S424–S435, doi:10.1086/667538.
33. Pigliucci, *Phenotypic Plasticity*.
34. T. J. DeWitt, A. Sih, and D. S. Wilson, "Costs and Limits of Phenotypic Plasticity," *Trends in Ecology & Evolution* 13 (1998): 77–81.
35. Jasienska, Nenko, and Jasienski, "Daughters Increase Longevity of Fathers."
36. K. Hawkes, J. F. O'Connell, and N. G. Blurton-Jones, "Hardworking Hadza Grandmothers," in *Comparative Socioecology*, ed. R. Foley (New York: Blackwell, 1989), 341–66.
37. Kuhnert and Nieschlag, "Reproductive Functions of the Ageing Male."
38. B. S. Hewlett and M. E. Lamb, *Hunter-Gatherer Childhoods: Evolutionary, Developmental, & Cultural Perspectives*, 1st ed. (New York: Aldine Transaction, 2005); M. Konner, *The Evolution of Childhood: Relationships, Emotion, Mind* (Cambridge, MA: Belknap Press of Harvard University Press, 2010).
39. M. Lahdenpera, A. F. Russell, and V. Lummaa, "Selection for Long Lifespan in Men: Benefits of Grandfathering?" *Proceedings of the Royal Society, Biological Sciences* 274 (2007): 2437–44, doi:10.1098/rspb.2007.0688.
40. C. S. Jamison, L. L. Cornell, P. L. Jamison, and H. Nakazato, "Are All Grandmothers Equal? A Review and a Preliminary Test of the 'Grandmother Hypothesis' in Tokugawa Japan," *American Journal of Physical Anthropology* 119 (2002): 67–76; A. Kemkes-Grottenthaler, "Of Grandmothers, Grandfathers and Wicked Step-Grandparents: Differential Impact of Paternal Grandparents on Grandoffspring Survival," *Historical Social Research/Historische Sozialforschung* 30 (2005):219–39.
41. Kuhnert and Nieschlag, "Reproductive Functions of the Ageing Male"; Tuljapurkar, Puleston, and Gurven, "Why Men Matter."
42. R. Lee, "Sociality, Selection, and Survival: Simulated Evolution of Mortality with Intergenerational Transfers and Food Sharing," *Proceedings of the National Academy of Sciences of the United States of America* 105 (2008): 7124–28, doi:0710234105 [pii], 10.1073/pnas.0710234105.
43. Hill and Hurtado, *Ache Life History*.
44. S. Beckerman and P. Valentine, *Cultures of Multiple Fathers: The Theory and Practice of Partible Paternity in Lowland South America* (Gainesville: University Press of Florida, 2002); R. S. Walker, M. V. Flinn, and K. R. Hill, "Evolutionary History of Partible Paternity in Lowland South America," *Proceedings of the National Academy of Sciences* 107 (2010): 19195–200, doi:1002598107 [pii], 10.1073/pnas.1002598107.
45. C. T. Palmer, "The Use and Abuse of Darwinian Psychology: Its Impact on Attempts to Determine the Evolutionary Basis of Human Rape," *Ethology and Sociobiology* 13 (1992): 289–99; B. Smuts, "Male Aggression against Women: An Evolutionary Perspective," *Human Nature* 3 (1992): 1–44; R. Thornhill and N. W.

Thornhill, "Human Rape: An Evolutionary Analysis," *Ethology and Sociobiology* 4 (1983): 137–73.

CHAPTER 6: DARWINIAN HEALTH AND OTHER CONTRADICTIONS

1. R. M. Nesse et al., "Evolution in Health and Medicine Sackler Colloquium: Making Evolutionary Biology a Basic Science for Medicine," *Proceedings of the National Academy of Sciences of the United States of America* 107 Suppl. 1 (2010): 1800–1807, doi:10.1073/pnas.0906224106.
2. R. G. Bribiescas and P. T. Ellison, "How Hormones Mediate Trade-Offs in Human Health and Disease," in *Evolution in Health and Disease*, ed. S. C. Stearns and J. C. Koella, 77–93 (New York: Oxford University Press, 2007).
3. P. T. Ellison, "Evolutionary Tradeoffs," *Evolution, Medicine, and Public Health* (2014), doi:10.1093/emph/eou015.
4. C. Foresta et al., "Role of Zinc Trafficking in Male Fertility: From Germ to Sperm," *Human Reproduction* 29 (2014): 1134–45, doi:10.1093/humrep/deu075.
5. S. C. Stearns and J. C. Koella, *Evolution in Health and Disease*, 2nd ed. (New York: Oxford University Press, 2008).
6. Bribiescas and Ellison, "How Hormones Mediate Trade-Offs in Human Health and Disease."
7. Jasienska, Nenko, and Jasienski, "Daughters Increase Longevity of Fathers"; Ziomkiewicz et al., "Evidence for the Cost of Reproduction in Humans."
8. S. D. Bianco and U. B. Kaiser, "The Genetic and Molecular Basis of Idiopathic Hypogonadotropic Hypogonadism," *Nature Reviews, Endocrinology* 5 (2009): 569–76, doi:10.1038/nrendo.2009.177; J. Waldstreicher et al., "The Genetic and Clinical Heterogeneity of Gonadotropin-Releasing Hormone Deficiency in the Human," *Journal of Clinical Endocrinology & Metabolism* 81 (1996): 4388–95.
9. Waldstreicher et al., "The Genetic and Clinical Heterogeneity of Gonadotropin-Releasing Hormone Deficiency in the Human."
10. R. M. Sapolsky, *The Trouble with Testosterone: And Other Essays on the Biology of the Human Predicament* (New York: Scribner, 1997).
11. J. D. Wilson and C. Roehrborn, "Long-Term Consequences of Castration in Men: Lessons from the Skoptzy and the Eunuchs of the Chinese and Ottoman Courts," *Journal of Clinical Endocrinology & Metabolism* 84 (1999): 4324–31.
12. K. J. Min, C. K. Lee, and H. N. Park, "The Lifespan of Korean Eunuchs," *Current Biology* 22 (2012): R792–93, doi:10.1016/j.cub.2012.06.036.
13. Ketterson et al., "Testosterone and Avian Life Histories"; J. C. Wingfield, R. E. Hegner, A. M. Dufty Jr., and G. F. Ball, "The 'Challenge Hypothesis': Theoretical Implications for Patterns of Testosterone Secretion, Mating Systems, and Breeding Strategies," *American Naturalist* 136 (1990): 829–46.
14. Wingfield et al., "The 'Challenge Hypothesis.' "

15. M. Muller and R. W. Wrangham, "Dominance, Aggression, and Testosterone in Wild Chimpanzees: A Test of the 'Challenge Hypothesis,'" *Animal Behavior* 67 (2004): 113–23.
16. B. C. Trumble et al., "Physical Competition Increases Testosterone among Amazonian Forager-Horticulturalists: A Test of the 'Challenge Hypothesis,'" *Proceedings of the Royal Society, Biological Sciences* 279 (2012): 2907–12, doi:10.1098/rspb.2012.0455.
17. L. W. Tsai and R. M. Sapolsky, "Rapid Stimulatory Effects of Testosterone upon Myotubule Metabolism and Sugar Transport, as Assessed by Silicon Microphysiometry," *Aggressive Behavior* 22 (1996): 357–64.
18. A. B. Araujo et al., "Prevalence of Symptomatic Androgen Deficiency in Men," *Journal of Clinical Endocrinology & Metabolism* 92, no. 11 (2007): 4241–47.
19. W. D. Finkle et al., "Increased Risk of Non-fatal Myocardial Infarction Following Testosterone Therapy Prescription in Men," *PLoS One* 9 (2014): e85805, doi: 10.1371/journal.pone.0085805.
20. M. P. Muehlenbein, J. Alger, F. Cogswell, M. James, and D. Krogstad, "The Reproductive Endocrine Response to Plasmodium Vivax Infection in Hondurans," *American Journal of Tropical Medicine and Hygiene* 73 (2005): 178–87.
21. Maklakov and Lummaa, "Evolution of Sex Differences in Lifespan and Aging."
22. Jasienska, *The Fragile Wisdom*; Eaton et al., "Women's Reproductive Cancers in Evolutionary Context"; W. Trevathan, E. O. Smith, and J. J. McKenna, *Evolutionary Medicine and Health: New Perspectives* (New York: Oxford University Press, 2008).
23. E. Kreiter, A. Richardson, J. Potter, and Y. Yasui, "Breast Cancer: Trends in International Incidence in Men and Women," *British Journal of Cancer* 110 (2014): 1891–97, doi:10.1038/bjc.2014.66.
24. S. C. Baca et al., "Punctuated Evolution of Prostate Cancer Genomes," *Cell* 153 (2013): 666–77, doi:10.1016/j.cell.2013.03.021.
25. A. Jemal et al., "Global Cancer Statistics," *CA: A Cancer Journal for Clinicians* 61 (2011): 69–90, doi:10.3322/caac.20107.
26. L. T. Amundadottir et al., "A Common Variant Associated with Prostate Cancer in European and African Populations," *Nature Genetics* 38 (2006): 652–58, doi: 10.1038/ng1808.
27. Jemal et al., "Global Cancer Statistics."
28. B. C. Trumble et al., "Challenging the Inevitability of Prostate Enlargement: Low Levels of Benign Prostatic Hyperplasia among Tsimane Forager-Horticulturalists," *Journals of Gerontology Series A: Biological Sciences and Medical Sciences* 70 (2015): 1262–68, doi:10.1093/gerona/glv051.
29. B. Campbell, "High Rate of Prostate Symptoms among Ariaal Men from Northern Kenya," *Prostate* 62 (2005): 83–90, doi:10.1002/pros.20120.
30. L. Calistro Alvarado, "Population Differences in the Testosterone Levels of Young Men Are Associated with Prostate Cancer Disparities in Older Men," *American Journal of Human Biology* 22 (2010): 449–55, doi:10.1002/ajhb.21016.

31. K. Ito, "Prostate Cancer in Asian Men," *Nature Reviews Urology* 11 (2014): 197–212, doi:10.1038/nrurol.2014.42.
32. R. G. Bribiescas, *Men: Evolutionary and Life History* (Cambridge, MA: Harvard University Press, 2006).
33. A. Hill, S. Ward, A. Deino, G. Curtis, and R. Drake, "Earliest Homo," *Nature* 355 (1992): 719–22, doi:10.1038/355719a0.
34. F. Buena et al., "Sexual Function Does Not Change When Serum Testosterone Levels Are Pharmacologically Varied within the Normal Male Range," *Fertility and Sterility* 59 (1993): 1118–23.
35. Ellison et al., "Population Variation in Age-Related Decline in Male Salivary Testosterone."
36. G. F. Gonzales, L. Villaorduna, M. Gasco, J. Rubio, and C. Gonzales, "Maca (Lepidium meyenii Walp): Una revisión sobre sus propiedades biológicas [Maca (Lepidium Meyenii Walp): A Review of Its Biological Properties]," *Revista Peruana de Medicina Experimental y Salud Pública* 31 (2014): 100–110; M. Oshima, Y. Gu, and S. Tsukada, "Effects of Lepidium Meyenii Walp and Jatropha macrantha on Blood Levels of Estradiol-17 Beta, Progesterone, Testosterone and the Rate of Embryo Implantation in Mice," *Journal of Veterinary Medical Science* 65 (2003): 1145–46.
37. Y. Oi et al., "Garlic Supplementation Increases Testicular Testosterone and Decreases Plasma Corticosterone in Rats Fed a High Protein Diet," *Journal of Nutrition* 131 (2001): 2150–56.
38. J. Prins, M. H. Blanker, A. M. Bohnen, S. Thomas, and J. L. Bosch, "Prevalence of Erectile Dysfunction: A Systematic Review of Population-Based Studies," *International Journal of Impotence Research* 14 (2002): 422–32.
39. P. Gray and B. Campbell, "Erectile Dysfunction and Its Correlates among the Ariaal of Northern Kenya," *International Journal of Impotence Research* 17 (2005): 445–49.
40. E. Becher and S. Glina, "Erectile Dysfunction in Latin America and Treatment with Sildenafil Citrate (Viagra): Introduction," *International Journal of Impotence Research* 14 Suppl. 2 (2002): S1–2, doi:10.1038/sj.ijir.3900889; D. J. Carbone Jr. and A. D. Seftel, "Erectile Dysfunction: Diagnosis and Treatment in Older Men," *Geriatrics* 57 (2002): 18–24; A. Imai et al., "Risk Factors for Erectile Dysfunction in Healthy Japanese Men," *International Journal of Andrology* 33 (2010): 569–73, doi:10.1111/j.1365-2605.2009.00974.x; A. I. Olugbenga-Bello, O. A. Adeoye, A. A. Adeomi, and A. O. Olajide, "Prevalence of Erectile Dysfunction (ED) and Its Risk Factors among Adult Men in a Nigerian Community," *Nigerian Postgraduate Medical Journal* 20 (2013): 130–35.
41. Prins et al., "Prevalence of Erectile Dysfunction."
42. R. D. Nadler and E. S. Bartlett, "Penile Erection: A Reflection of Sexual Arousal and Arousability in Male Chimpanzees," *Physiology & Behavior* 61 (1997): 425–32.

43. A. A. Sandel, "Brief Communication: Hair Density and Body Mass in Mammals and the Evolution of Human Hairlessness," *American Journal of Physical Anthropology* 152 (2013): 145–50, doi:10.1002/ajpa.22333.
44. J. M. Wood et al., "Senile Hair Graying: H2O2-Mediated Oxidative Stress Affects Human Hair Color by Blunting Methionine Sulfoxide Repair," *FASEB Journal* 23 (2009): 2065–75, doi:10.1096/fj.08-125435.
45. M. N. Muller, M. E. Thompson, and R. W. Wrangham, "Male Chimpanzees Prefer Mating with Old Females," *Current Biology* 16 (2006): 2234–38, doi:10.1016/j.cub.2006.09.042.
46. B. J. Dixson and P. L. Vasey, "Beards Augment Perceptions of Men's Age, Social Status, and Aggressiveness, But Not Attractiveness," *Behavioral Ecology* 23 (2012): 481–90, doi:10.1093/beheco/arr214.
47. S. Christiansen et al., "Synergistic Disruption of External Male Sex Organ Development by a Mixture of Four Antiandrogens," *Environmental Health Perspectives* 117 (2009): 1839–46, doi:10.1289/ehp.0900689.
48. M. A. Novak and J. S. Meyer, "Alopecia: Possible Causes and Treatments, Particularly in Captive Nonhuman Primates," *Comparative Medicine* 59 (2009): 18–26.
49. R. J. Fredericksen, "Just Kill Me When I'm 50: Impact of Gay American Culture on Young Gay Men's Perceptions of Aging," *Anthropology & Aging Quarterly* 31 (2010): 27–35; D. C. Kimmel, T. Rose, and S. David, *Lesbian, Gay, Bisexual, and Transgender Aging: Research and Clinical Perspectives* (New York: Columbia University Press, 2006); T. Witten and A. E. Eyler, *Gay, Lesbian, Bisexual, & Transgender Aging: Challenges in Research, Practice, and Policy* (Baltimore: Johns Hopkins University Press, 2012); J. M. Cruz, *Sociological Analysis of Aging: The Gay Male Perspective* (New York: Harrington Park Press, 2003).
50. R. C. Kirkpatrick, "The Evolution of Human Homosexual Behavior," *Current Anthropology* 41 (2000): 385–413.
51. E. Abraham et al., "Father's Brain Is Sensitive to Childcare Experiences," *Proceedings of the National Academy of Sciences* 111, no. 27 (2014): 9792–97, doi:10.1073/pnas.1402569111.
52. S. F. Anestis, "Female Genito-Genital Rubbing in a Group of Captive Chimpanzees," *International Journal of Primatology* 25 (2004): 477–88.
53. R. D. Schope, "Who's Afraid of Growing Old? Gay and Lesbian Perceptions of Aging," *Journal of Gerontological Social Work* 45 (2005): 23–38, doi:10.1300/J083v45n04_03.
54. P. Cameron, K. Cameron, and W. L. Playfair, "Does Homosexual Activity Shorten Life?" *Psychological Reports* 83 (1998): 847–66, doi:10.2466/pr0.1998.83.3.847.
55. Although the constitutionality of same-sex marriage has only recently been upheld by the United States Supreme Court, it has been legal since 1989 in Denmark.
56. M. Frisch and H. Bronnum-Hansen, "Mortality among Men and Women in Same-Sex Marriage: A National Cohort Study of 8333 Danes," *American Journal of Public Health* 99 (2009): 133–37, doi:10.2105/AJPH.2008.133801.

57. P. N. Halkitis et al., "Evidence for a Syndemic in Aging HIV-Positive Gay, Bisexual, and Other MSM: Implications for a Holistic Approach to Prevention and Healthcare," *Natural Resource Modeling* 36 (2012), doi:10.1111/napa.12009.
58. There are a few exceptions but many, understandably, focus on the impact of AIDS and HIV health issues among the gay community. P. N. Halkitis, F. Kapadia, D. C. Ompad, and R. Perez-Figueroa, "Moving toward a Holistic Conceptual Framework for Understanding Healthy Aging among Gay Men," *Journal of Homosexuality* 62, no. 5 (2015): 571–87.
59. R. G. Wight, F. Harig, C. S. Aneshensel, and R. Detels, "Depressive Symptom Trajectories, Aging-Related Stress, and Sexual Minority Stress among Midlife and Older Gay Men: Linking Past and Present," *Research on Aging* (2015), doi:10.1177/0164027515590423.
60. R. G. Wight, A. J. LeBlanc, B. de Vries, and R. Detels, "Stress and Mental Health among Midlife and Older Gay-Identified Men," *American Journal of Public Health* 102 (2012): 503–10, doi:10.2105/AJPH.2011.300384.

CHAPTER 7: OLDER MEN AND THE FUTURE OF HUMAN EVOLUTION

1. http://www.gallup.com/poll/21814/evolution-creationism-intelligent-design.aspx.
2. The night sky around the equator is filled with satellites that are easy to see outside of city lights.
3. R. S. Walker, M. J. Hamilton, and A. A. Groth, "Remote Sensing and Conservation of Isolated Indigenous Villages in Amazonia," *Royal Society Open Science* 1 (2014), doi:10.1098/rsos.140246.
4. R. S. Walker and K. R. Hill, "Protecting Isolated Tribes," *Science* 348 (2015): 1061, doi:10.1126/science.aac6540.
5. I have yet to see an older forager woman with eyeglasses, which perhaps reveals evidence of a sexual disparity in the allocation or resources. Or it may be that women maintain better eyesight with age compared to men. At the moment, this observation is anecdotal and needs further investigation.
6. Goldstein, "A Secular Trend toward Earlier Male Sexual Maturity"; Kruger and Nesse, "Sexual Selection and the Male."
7. Y. Choi, M. S. Fabic, S. Hounton, and D. Koroma, "Meeting Demand for Family Planning within a Generation: Prospects and Implications at Country Level," *Global Health Action* 8 (2015), doi:10.3402/gha.v8.29734.
8. Evolutionary biologist Ernst Mayr once stated that the term "natural selection" was problematic since it implied that there was "unnatural or artificial selection" at the hands of humans. This placed humans outside of the natural world, which is a stance that Mayer did not agree with. He stated that natural selection could be more accurately described as nonrandom elimination. That is, certain individuals will leave future descendants and genes behind as a result of environmental challenges.

9. The evolution of warfare has sometimes also included discussion of ants and other eusocial insects. However, the behavioral genetics of these organisms is quite different from that of mammals and probably should not be included with the types of organized conflicts seen in humans and chimpanzees. R. W. Wrangham, "Evolution of Coalitionary Killing," *American Journal of Physical Anthropology* 42 Suppl. 29 (1999): 1–30.
10. S. Pinker, *The Better Angels of Our Nature: Why Violence Has Declined* (New York: Viking, 2011).
11. Hill and Hurtado, *Ache Life History.*
12. S. L. Rubenstein, "Circulation, Accumulation, and the Power of Shuar Shrunken Heads," *Cultural Anthropology* 22 (2007): 357–99, doi:10.1525/Can.2007.22.3.357.
13. http://dialogo-americas.com/en_GB/articles/rmisa/features/regional_news/2012/03/20/feature-ex-2970.
14. A. Gray, D. N. Jackson, and J. B. McKinlay, "The Relation between Dominance, Anger, and Hormones in Normally Aging Men: Results from the Massachusetts Male Aging Study," *Psychosomatic Medicine* 53 (1991): 375–85.
15. L. Glowacki and R. Wrangham, "Warfare and Reproductive Success in a Tribal Population," *Proceedings of the National Academy of Sciences of the United States of America* 112 (2015): 348–53, doi:10.1073/pnas.1412287112.
16. Among the Turkana of Kenya, stealth raids involve quick attacks with only a few men. Battle raids are more overt attacks on enemies that involve more men and result in higher mortality.
17. M. R. Zefferman, R. Baldini, and S. Mathew, "Solving the Puzzle of Human Warfare Requires an Explanation of Battle Raids and Cultural Institutions," *Proceedings of the National Academy of Sciences of the United States of America* 112 (2015): E2557, doi:10.1073/pnas.1504458112.
18. R. G. Bribiescas, "An Evolutionary and Life History Perspective on the Role of Males on Human Futures," *Futures* 43 (2011): 729–39, doi:http://dx.doi.org/10.1016/j.futures.2011.05.016.
19. Ibid.
20. Wrangham and Peterson, *Demonic Males.*
21. R. G. Bribiescas and T. Gendler, "An Excellent Faculty Is a Diverse Faculty," in *Yale Daily News* (New Haven, CT, 2015).
22. C. A. Moss-Racusin, J. F. Dovidio, V. L. Brescoll, M. J. Graham, and J. Handelsman, "Science Faculty's Subtle Gender Biases Favor Male Students," *Proceedings of the National Academy of Sciences of the United States of America* 109 (2012): 16474–79, doi:10.1073/pnas.1211286109; S. L. Rose et al., "Gender Differences in Physicians' Financial Ties to Industry: A Study of National Disclosure Data," *PLoS One* 10 (2015): e0129197, doi:10.1371/journal.pone.0129197.
23. It should be noted that while older men are often the decision makers for appointments and promotions in academia, especially within the STEM (science, technology, engineering, mathematics) fields, research shows that women also exhibit

unconscious bias against other women when they are involved in the selection/promotion process. Older men exert their influence by their sheer number in these processes.

Leadership in academia is still predominantly male. Women have made considerable strides in leadership at Yale, but the university has yet to appoint a woman president after over 310 years of existence. Similarly it took Harvard 350 years to appoint a woman president. For Oxford and Cambridge, it took over 800 years before a woman was appointed to the helm.

INDEX